Springer Actuarial
Lecture Notes

This is a series on actuarial topics in a broad and interdisciplinary sense, aimed at students, academics and practitioners in the fields of insurance and finance.

Springer Actuarial informs timely on theoretical and practical aspects of topics like risk management, internal models, solvency, asset-liability management, market-consistent valuation, the actuarial control cycle, insurance and financial mathematics, and other related interdisciplinary areas.

The series aims to serve as a primary scientific reference for education, research, development and model validation.

The type of material considered for publication includes lecture notes, monographs and textbooks. All submissions will be peer-reviewed.

More information about this series at http://www.springer.com/series/15682

Hanspeter Schmidli

Risk Theory

Springer

Hanspeter Schmidli
Institute of Mathematics
University of Cologne
Germany

Springer Actuarial
Lecture Notes
ISBN 978-3-319-72004-3 ISBN 978-3-319-72005-0 (eBook)
https://doi.org/10.1007/978-3-319-72005-0

Library of Congress Control Number: 2017962068

Mathematics Subject Classification (2010): 91B30, 60F10, 60G42, 60G44, 60G51, 60G55, 60J10, 60J25, 60K05, 60K20, 91B16

Printed on acid-free paper

This Springer imprint is published by the registered company Springer International Publishing AG part of Springer Nature
The registered company address is: Gewerbestrasse 11, 6330 Cham, Switzerland

Preface

This book aims to give an introduction to the methods used in non-life insurance. In addition to providing an overview of classical actuarial methods, the main part deals with ruin models, which is particularly interesting from a mathematical point of view. However, ruin theory also gives a deeper insight into and understanding how losses or an insolvency happens, if it happens at all, and what precautions may be taken to avoid an undesirable situation. Even though "ruin in infinite time" is not considered by the solvency rules, the theory gives an understanding of the risks taken.

I started writing this book back in 1994, when I gave a lecture on risk theory at Heriot–Watt University in Edinburgh. The lecture was based on notes by my colleagues, in particular by Howard Waters. The main part of the book has its origins in a two-semester course I gave at the University of Aarhus from 1994 to 2000. Parts of the notes to this course were also used in the book project [110]. In the last year, I added the chapter on claims reserving and the discussions on some aspects of solvency. I hope that all the theoretical background that an actuary may need can now be found in this book.

Many colleagues have directly or indirectly contributed to this book. I want to thank my Ph.D. supervisor Paul Embrechts, who started my interest in this topic; my former colleagues in Edinburgh, in particular Howard Waters, who supported me during my stay in Scotland's capital; Jan Grandell and Søren Asmussen, whose experience has increased my research skills; Hansjörg Albrecher, who drew my attention to several interesting references; and Mario Wüthrich, whose fruitful discussions with me on claims reserving has improved the corresponding chapter. Last but not least, I thank my wife, who has always supported my scientific career, even though it was often quite hard for her, in particular, when the children were small.

Cologne, Germany
September 2017

Hanspeter Schmidli

Contents

Principal Notation

$\mathbb{I}_A$	Indicator function
$\mathbb{E}$	Expected value
$\mathbb{N}$	The natural numbers, $\mathbb{N} = \{0, 1, 2, \ldots\}$
$\mathbb{P}$	The basic probability measure
$\mathbb{R}$	The real numbers
$\mathbb{R}_+$	The positive real numbers, $\mathbb{R}_+ = [0, \infty)$
$\mathbb{Z}$	The integers, $\mathbb{Z} = \{\ldots, -2, -1, 0, 1, 2, \ldots\}$
$\mathcal{F}$	σ-algebra of the probability space
$\{\mathcal{F}_t\}$	Filtration
$\{\mathcal{F}_t^X\}$	Natural filtration of the process X
$\mathcal{L}^\infty$	Space of bounded random variables
$\{C_t\}$	Surplus process
c	Premium rate
F	Distribution function or interclaim arrival distribution
G	Claim size distribution
$M_Y(r)$	Moment-generating function of the random variable Y
$N(= \{N_t\})$	Claim number or claim number process
R	Lundberg coefficient
T_i	ith occurrence time
$x^+ = x \vee 0$	Positive part of x
$x^- = (-x) \vee 0$	Negative part of x
$\lfloor x \rfloor$	Integer part of x, $\lfloor x \rfloor = \sup\{n \in \mathbb{Z} : n \leq x\}$
Y_k	Claim sizes
Ω	Event space on which probabilities are defined
$\delta(x)$	Survival probability
Φ	Standard normal distribution function
λ	Claim arrival intensity
μ, μ_n	nth moment of the claim sizes

$\psi(x)$	Ruin probability
τ	Time of ruin
Θ_i	Risk characteristics
$\theta(r)$	Cumulant-generating function of the surplus

Chapter 1
Risk Models

In this chapter we will consider a risk in a single time period. We will see how to approximate the distribution of a compound sum, how to calculate premia, and we will introduce risk measures.

1.1 Introduction

Let us consider a (collective) insurance contract in some fixed time period $(0, T]$, for instance $T = 1$ year. Let N denote the number of claims in $(0, T]$ and $Y_1, Y_2, \ldots, Y_N$ the corresponding claims. Then

$$S = \sum_{i=1}^{N} Y_i$$

is the accumulated sum of claims. We assume

(i) N and $\{Y_1, Y_2, \ldots\}$ are independent.
(ii) $Y_1, Y_2, \ldots$ are independent.
(iii) $Y_1, Y_2, \ldots$ have the same distribution function, G say.

We further assume that $G(0) = 0$, i.e. the claim amounts are positive. Let $M_Y(r) = \mathbb{E}[e^{rY_i}]$, $\mu_n = \mathbb{E}[Y_1^n]$ if the expressions exist and $\mu = \mu_1$. The distribution of S can be written as

$$\begin{aligned}\mathbb{P}[S \le x] = \mathbb{E}[\mathbb{P}[S \le x \mid N]] &= \sum_{n=0}^{\infty} \mathbb{P}[S \le x \mid N = n]\mathbb{P}[N = n] \\ &= \sum_{n=0}^{\infty} \mathbb{P}[N = n] G^{*n}(x)\,.\end{aligned}$$

H. Schmidli, *Risk Theory*, Springer Actuarial,
DOI:10.1007/978-3-319-72005-0_1

In general, this is not easy to compute, but it is often enough to know only some characteristics of a distribution.

$$\mathbb{E}[S] = \mathbb{E}\Big[\sum_{i=1}^{N} Y_i\Big] = \mathbb{E}\Big[\mathbb{E}\Big[\sum_{i=1}^{N} Y_i \;\Big|\; N\Big]\Big] = \mathbb{E}\Big[\sum_{i=1}^{N} \mu\Big] = \mathbb{E}[N\mu] = \mathbb{E}[N]\mu$$

is called the first Wald formula, and

$$\begin{aligned}\mathbb{E}[S^2] &= \mathbb{E}\Big[\mathbb{E}\Big[\Big(\sum_{i=1}^{N} Y_i\Big)^2 \;\Big|\; N\Big]\Big] = \mathbb{E}\Big[\mathbb{E}\Big[\sum_{i=1}^{N}\sum_{j=1}^{N} Y_i Y_j \;\Big|\; N\Big]\Big] \\ &= \mathbb{E}[N\mu_2 + N(N-1)\mu^2] = \mathbb{E}[N^2]\mu^2 + \mathbb{E}[N](\mu_2 - \mu^2)\end{aligned}$$

gives the second Wald formula

$$\mathrm{Var}[S] = \mathrm{Var}[N]\mu^2 + \mathbb{E}[N]\mathrm{Var}[Y_1]\,.$$

The moment generating function of S becomes

$$\begin{aligned}M_S(r) &= \mathbb{E}[e^{rS}] = \mathbb{E}\Big[\exp\Big\{r\sum_{i=1}^{N} Y_i\Big\}\Big] = \mathbb{E}\Big[\prod_{i=1}^{N} e^{rY_i}\Big] = \mathbb{E}\Big[\mathbb{E}\Big[\prod_{i=1}^{N} e^{rY_i} \;\Big|\; N\Big]\Big] \\ &= \mathbb{E}\Big[\prod_{i=1}^{N} M_Y(r)\Big] = \mathbb{E}\big[(M_Y(r))^N\big] = \mathbb{E}\Big[e^{N\log(M_Y(r))}\Big] \\ &= M_N(\log(M_Y(r)))\,,\end{aligned}$$

where $M_N(r)$ is the moment generating function of N. The coefficient of skewness $\mathbb{E}[(S-\mathbb{E}[S])^3]/(\mathrm{Var}[S])^{3/2}$ can be calculated from the moment generating function by using the formula

$$\mathbb{E}[(S-\mathbb{E}[S])^3] = \frac{d^3}{dr^3}\log(M_S(r))\Big|_{r=0}\,.$$

1.2 The Compound Binomial Model

We model $N \sim \mathrm{B}(n, p)$ for some $n \in \mathbb{N}$ and $p \in (0, 1)$. We obtain

$$\mathbb{E}[S] = np\mu\,,$$

$$\mathrm{Var}[S] = np(1-p)\mu^2 + np(\mu_2 - \mu^2) = np(\mu_2 - p\mu^2)$$

and

$$M_S(r) = (pM_Y(r) + 1 - p)^n .$$

Let us consider another characteristic of the distribution of S, the skewness. We need to compute $\mathbb{E}[(S - \mathbb{E}[S])^3]$.

$$\begin{aligned}
\frac{\mathrm{d}^3}{\mathrm{d}r^3} n \log(pM_Y(r) + 1 - p) &= n\frac{\mathrm{d}^2}{\mathrm{d}r^2}\left(\frac{pM_Y'(r)}{pM_Y(r) + 1 - p}\right) \\
&= n\frac{\mathrm{d}}{\mathrm{d}r}\left(\frac{pM_Y''(r)}{pM_Y(r) + 1 - p} - \frac{p^2M_Y'(r)^2}{(pM_Y(r) + 1 - p)^2}\right) \\
&= n\left(\frac{pM_Y'''(r)}{pM_Y(r) + 1 - p} - \frac{3p^2M_Y''(r)M_Y'(r)}{(pM_Y(r) + 1 - p)^2} + \frac{2p^3(M_Y'(r))^3}{(pM_Y(r) + 1 - p)^3}\right) .
\end{aligned}$$

For $r = 0$ we get

$$\mathbb{E}[(S - \mathbb{E}[S])^3] = n(p\mu_3 - 3p^2\mu_2\mu + 2p^3\mu^3) ,$$

from which the coefficient of skewness can be calculated.

Example 1.1 Assume that the claim amounts are deterministic, y_0 say. Then

$$\mathbb{E}[(S - \mathbb{E}[S])^3] = ny_0^3(p - 3p^2 + 2p^3) = 2ny_0^3p(\frac{1}{2} - p)(1 - p) .$$

Thus

$$\mathbb{E}[(S - \mathbb{E}[S])^3] \gtreqless 0 \iff p \lesseqgtr \frac{1}{2} .$$

■

1.3 The Compound Poisson Model

In addition to the compound binomial model we assume that n is large and p is small. Let $\lambda = np$. Because

$$B(n, \lambda/n) \longrightarrow \text{Pois}(\lambda) \qquad \text{as } n \to \infty$$

it is natural to model

$$N \sim \text{Pois}(\lambda) .$$

We get

$$\mathbb{E}[S] = \lambda\mu ,$$

$$\text{Var}[S] = \lambda\mu^2 + \lambda(\mu_2 - \mu^2) = \lambda\mu_2$$

and

$$M_S(r) = \exp\{\lambda(M_Y(r) - 1)\}\,.$$

Let us also compute the coefficient of skewness.

$$\frac{\mathrm{d}^3}{\mathrm{d}r^3}\log(M_S(r)) = \frac{\mathrm{d}^3}{\mathrm{d}r^3}(\lambda(M_Y(r) - 1)) = \lambda M_Y'''(r)$$

and thus

$$\mathbb{E}[(S - \mathbb{E}[S])^3] = \lambda\mu_3\,.$$

The coefficient of skewness is

$$\frac{\mathbb{E}[(S - \mathbb{E}[S])^3]}{(\mathrm{Var}[S])^{3/2}} = \frac{\mu_3}{\sqrt{\lambda\mu_2^3}} > 0\,.$$

Problem: The distribution of S is always positively skewed.

Example 1.2 An insurance company models claims from fire insurance as $\mathrm{LN}(m, \sigma^2)$. Let us first compute the moments

$$\mu_n = \mathbb{E}[Y_1^n] = \mathbb{E}[\mathrm{e}^{n\log Y_1}] = M_{\log Y_1}(n) = \exp\{\sigma^2 n^2/2 + nm\}\,.$$

Thus

$$\mathbb{E}[S] = \lambda\exp\{\sigma^2/2 + m\}\,,$$

$$\mathrm{Var}[S] = \lambda\exp\{2\sigma^2 + 2m\}$$

and

$$\frac{\mathbb{E}[(S - \mathbb{E}[S])^3]}{(\mathrm{Var}[S])^{3/2}} = \frac{\exp\{9\sigma^2/2 + 3m\}}{\sqrt{\lambda\exp\{6\sigma^2 + 6m\}}} = \frac{\exp\{3\sigma^2/2\}}{\sqrt{\lambda}}\,.$$

■

The computation of the characteristics of a risk is much easier for the compound Poisson model than for the compound binomial model. Using a compound Poisson model also has another great advantage. Assume that a portfolio consists of several independent single risks $S^{(1)}, S^{(2)}, \ldots, S^{(j)}$, each modelled as compound Poisson. For simplicity we use $j = 2$ in the following calculation. We want to find the moment generating function of $S^{(1)} + S^{(2)}$.

$$\begin{aligned} M_{S^{(1)}+S^{(2)}}(r) &= M_{S^{(1)}}(r)M_{S^{(2)}}(r) \\ &= \exp\{\lambda^{(1)}(M_{Y^{(1)}}(r) - 1)\}\exp\{\lambda^{(2)}(M_{Y^{(2)}}(r) - 1)\} \\ &= \exp\Bigl\{\lambda\Bigl(\frac{\lambda^{(1)}}{\lambda}M_{Y^{(1)}}(r) + \frac{\lambda^{(2)}}{\lambda}M_{Y^{(2)}}(r) - 1\Bigr)\Bigr\}\,, \end{aligned}$$

where $\lambda = \lambda^{(1)} + \lambda^{(2)}$. It follows that $S^{(1)} + S^{(2)}$ is compound Poisson distributed with Poisson parameter λ and claim size distribution function

$$G(x) = \frac{\lambda^{(1)}}{\lambda} G^{(1)}(x) + \frac{\lambda^{(2)}}{\lambda} G^{(2)}(x) .$$

The claim size can be obtained by choosing it from the first risk with probability $\lambda^{(1)}/\lambda$ and from the second risk with probability $\lambda^{(2)}/\lambda$.

Let us now split the claim amounts into different classes. Choose some disjoint sets $A_1, A_2, \ldots, A_m$ with $\mathbb{P}[Y_1 \in \cup_{k=1}^m A_k] = 1$. Let $p_k = \mathbb{P}[Y_1 \in A_k]$ be the probability that a claim is in claim size class k. We can assume that $p_k > 0$ for all k. We denote by N_k the number of claims in claim size class k. Because the claim amounts are independent it follows that, given $N = n$, the vector $(N_1, N_2, \ldots, N_m)$ is conditionally multinomial distributed with parameters $n, p_1, p_2, \ldots, p_m$. We now want to find the unconditioned distribution of $(N_1, N_2, \ldots, N_m)$. Let $n_1, n_2, \ldots, n_m$ be natural numbers and $n = n_1 + n_2 + \cdots + n_m$.

$$\begin{aligned}
&\mathbb{P}[N_1 = n_1, N_2 = n_2, \ldots, N_m = n_m] \\
&\quad = \mathbb{P}[N_1 = n_1, N_2 = n_2, \ldots, N_m = n_m, N = n] \\
&\quad = \mathbb{P}[N_1 = n_1, N_2 = n_2, \ldots, N_m = n_m \mid N = n]\mathbb{P}[N = n] \\
&\quad = \frac{n!}{n_1! n_2! \cdots n_m!} p_1^{n_1} p_2^{n_2} \cdots p_m^{n_m} \frac{\lambda^n}{n!} \mathrm{e}^{-\lambda} = \prod_{k=1}^m \frac{(\lambda p_k)^{n_k}}{n_k!} \mathrm{e}^{-\lambda p_k} .
\end{aligned}$$

It follows that $N_1, N_2, \ldots, N_m$ are independent and that N_k is Pois(λp_k) distributed. Because the claim sizes are independent of N the risks

$$S_k = \sum_{i=1}^N Y_i \mathbb{1}_{\{Y_i \in A_k\}}$$

are compound Poisson distributed with Poisson parameter λp_k and claim size distribution

$$G_k(x) = \mathbb{P}[Y_1 \le x \mid Y_1 \in A_k] = \frac{\mathbb{P}[Y_1 \le x, Y_1 \in A_k]}{\mathbb{P}[Y_1 \in A_k]} .$$

Moreover, the sums $\{S_k\}$ are independent. In the special case where $A_k = (t_{k-1}, t_k]$ we get

$$G_k(x) = \frac{G(x) - G(t_{k-1})}{G(t_k) - G(t_{k-1})} \qquad (t_{k-1} \le x \le t_k).$$

1.4 The Compound Mixed Poisson Model

As mentioned before it is a disadvantage of the compound Poisson model that the distribution is always positively skewed. In practice, it also often turns out that the model does not allow enough fluctuations. For instance, $\mathbb{E}[N] = \text{Var}[N]$. A simple way to allow for more fluctuation is to let the parameter λ be stochastic. Let H denote the distribution function of λ.

$$\mathbb{P}[N = n] = \mathbb{E}[\mathbb{P}[N = n \mid \lambda]] = \mathbb{E}\Big[\frac{\lambda^n}{n!}\mathrm{e}^{-\lambda}\Big] = \int_0^\infty \frac{\ell^n}{n!}\mathrm{e}^{-\ell}\,\mathrm{d}H(\ell)\,.$$

The moments are

$$\mathbb{E}[S] = \mathbb{E}[\mathbb{E}[S \mid \lambda]] = \mathbb{E}[\lambda\mu] = \mathbb{E}[\lambda]\mu\,, \tag{1.1a}$$

$$\mathbb{E}[S^2] = \mathbb{E}[\mathbb{E}[S^2 \mid \lambda]] = \mathbb{E}[\lambda\mu_2 + \lambda^2\mu^2] = \mathbb{E}[\lambda^2]\mu^2 + \mathbb{E}[\lambda]\mu_2 \tag{1.1b}$$

and

$$\begin{aligned}\mathbb{E}[S^3] &= \mathbb{E}[\mathbb{E}[(\{S - \lambda\mu\} + \lambda\mu)^3 \mid \lambda]] \\ &= \mathbb{E}[\lambda^3]\mu^3 + 3\mathbb{E}[\lambda^2]\mu_2\mu + \mathbb{E}[\lambda]\mu_3\,.\end{aligned} \tag{1.1c}$$

Thus the variance is

$$\text{Var}[S] = \text{Var}[\lambda]\mu^2 + \mathbb{E}[\lambda]\mu_2$$

and the third centralised moment becomes

$$\begin{aligned}\mathbb{E}[(S - \mathbb{E}[\lambda]\mu)^3] &= \mathbb{E}[\lambda^3]\mu^3 + 3\mathbb{E}[\lambda^2]\mu_2\mu + \mathbb{E}[\lambda]\mu_3 \\ &\quad -3(\mathbb{E}[\lambda^2]\mu^2 + \mathbb{E}[\lambda]\mu_2)\mathbb{E}[\lambda]\mu + 2\mathbb{E}[\lambda]^3\mu^3 \\ &= \mathbb{E}[(\lambda - \mathbb{E}[\lambda])^3]\mu^3 + 3\text{Var}[\lambda]\mu_2\mu + \mathbb{E}[\lambda]\mu_3\,.\end{aligned}$$

We can see that the coefficient of skewness can also be negative.

It remains to compute the moment generating function

$$M_S(r) = \mathbb{E}\left[\mathbb{E}\left[\mathrm{e}^{rS} \mid \lambda\right]\right] = \mathbb{E}[\exp\{\lambda(M_Y(r) - 1)\}] = M_\lambda(M_Y(r) - 1)\,. \tag{1.2}$$

1.5 The Compound Negative Binomial Model

Let us first consider an example of the compound mixed Poisson model. Assume that $\lambda \sim \Gamma(\gamma, \beta)$. Then

$$M_N(r) = M_\lambda(\mathrm{e}^r - 1) = \Bigl(\frac{\beta}{\beta - (\mathrm{e}^r - 1)}\Bigr)^\gamma = \Bigl(\frac{\frac{\beta}{\beta+1}}{1 - (1 - \frac{\beta}{\beta+1})\mathrm{e}^r}\Bigr)^\gamma .$$

In this case N has a negative binomial distribution. Actuaries started using the negative binomial distribution for the number of claims a long time ago. They recognised that a Poisson distribution rarely fits the real data. Estimating the parameters in a negative binomial distribution yielded satisfactory results.

Let now $N \sim \mathrm{NB}(\alpha, p)$. Then

$$\mathbb{E}[S] = \frac{\alpha(1-p)}{p}\mu$$

and

$$\mathrm{Var}[S] = \frac{\alpha(1-p)}{p^2}\mu^2 + \frac{\alpha(1-p)}{p}(\mu_2 - \mu^2) .$$

The moment generating function is

$$M_S(r) = \Bigl(\frac{p}{1 - (1-p)M_Y(r)}\Bigr)^\alpha .$$

The third centralised moment can be computed as

$$\mathbb{E}[(S - \mathbb{E}[S])^3] = \alpha\Bigl(\frac{1}{p} - 1\Bigr)\mu_3 + 3\alpha\Bigl(\frac{1}{p} - 1\Bigr)^2\mu_2\mu + 2\alpha\Bigl(\frac{1}{p} - 1\Bigr)^3\mu^3 .$$

Note that the compound negative binomial distribution is always positively skewed. Thus the compound negative binomial distribution does not satisfy all the desired properties. Nevertheless, in practice almost all risks are positively skewed.

1.6 A Note on the Individual Model

Assume that a portfolio consist of m independent possibly not identically distributed individual contracts $(S^{(i)})_{i \le m}$. There can be at most one claim for each of the contracts. Such a claim occurs with probability $p^{(i)}$. Its size has distribution function $F^{(i)}$ and moment generating function $M^{(i)}(r)$. Let $\lambda = \sum_{i=1}^m p^{(i)}$. The moment generating function of the aggregate claims from the portfolio is

$$M_S(r) = \prod_{i=1}^m (1 + p^{(i)}(M^{(i)}(r) - 1)) .$$

The term $p^{(i)}(M^{(i)}(r) - 1)$ is small for r not too large. Consider the logarithm

$$\log(M_S(r)) = \sum_{i=1}^{m} \log(1 + p^{(i)}(M^{(i)}(r) - 1))$$
$$\approx \sum_{i=1}^{m} p^{(i)}(M^{(i)}(r) - 1) = \lambda\Bigl(\sum_{i=1}^{m} \frac{p^{(i)}}{\lambda}(M^{(i)}(r) - 1)\Bigr).$$

The last expression turns out to be the logarithm of the moment generating function of a compound Poisson distribution. In fact, this derivation is the reason why the compound Poisson model is very popular amongst actuaries.

The compound model can be used for a large, quite homogeneous model, where the number of claims is small compared to the number of contracts. If we imagine the possible claims from the different contracts are in an urn, and after a claim, we draw one of these claims, then it is very unlikely that we will draw the same claim twice.

1.7 A Note on Reinsurance

For most insurance companies the premium volume is not big enough to carry the complete risk. This is especially the case for large claim sizes, like insurance against damages caused by hurricanes or earthquakes. Therefore the insurers try to share part of the risk with other companies. Sharing the risk is done via reinsurance. Let S^I denote the part of the risk taken by the insurer and S^R the part taken by the reinsurer. Reinsurance can act on the individual claims or it can act on the whole risk S. Let f be an increasing function with $f(0) = 0$ and $f(x) \le x$ for all $x \ge 0$. A reinsurance form acting on the individual claims is

$$S^I = \sum_{i=1}^{N} f(Y_i)\,, \qquad S^R = S - S^I\,.$$

The most common reinsurance forms are

- proportional reinsurance $\quad f(x) = \alpha x, \quad (0 < \alpha < 1)$,
- excess of loss reinsurance $\quad f(x) = \min\{x, M\}, \quad (M > 0)$.

We will consider these two reinsurance forms in the sequel. A reinsurance form acting on the whole risk is

$$S^I = f(S)\,, \qquad S^R = S - S^I\,.$$

The most common example of this reinsurance form is the

- stop loss reinsurance $\quad f(x) = \min\{x, M\}, \quad (M > 0)$.

1.7.1 Proportional Reinsurance

For proportional reinsurance we have $S^I = \alpha S$, and thus

$$\mathbb{E}[S^I] = \alpha \mathbb{E}[S] ,$$
$$\mathrm{Var}[S^I] = \alpha^2 \mathrm{Var}[S] ,$$
$$\frac{\mathbb{E}[(S^I - \mathbb{E}[S^I])^3]}{(\mathrm{Var}[S^I])^{3/2}} = \frac{\mathbb{E}[(S - \mathbb{E}[S])^3]}{(\mathrm{Var}[S])^{3/2}}$$

and

$$M_{S^I}(r) = \mathbb{E}\left[\mathrm{e}^{r\alpha S}\right] = M_S(\alpha r) .$$

We can see that the coefficient of skewness does not change, but the variance is much smaller. The following considerations also suggest that the risk has decreased. Let the premium charged for this contract be p and assume that the insurer gets αp, the reinsurer $(1 - \alpha)p$. Let the initial capital of the company be u. The probability that the company gets ruined after one year is

$$\mathbb{P}[\alpha S > \alpha p + u] = \mathbb{P}[S > p + u/\alpha] .$$

The effect for the insurer is thus like having a larger initial capital.

Remark 1.3 The computations for the reinsurer can be obtained by replacing α by $(1 - \alpha)$. ■

1.7.2 Excess of Loss Reinsurance

Under excess of loss reinsurance we cannot give formulae of the above type for the cumulants and the moment generating function. They all have to be computed from the new distribution function of the claim sizes $Y_i^I = \min\{Y_i, M\}$. An indication that the risk has decreased for the insurer is that the claim sizes are bounded. This is wanted especially in the case of large claims.

For the calculation of the expected value of the payments, note that $\mathbb{P}[Y_i^R > x] = \mathbb{P}[Y > x + M] = 1 - G(x + M)$. Thus

$$\mathbb{E}[Y_i^R] = \int_0^\infty (1 - G(x + M))\, \mathrm{d}x = \int_M^\infty (1 - G(z))\, \mathrm{d}z .$$

In particular,

$$\mathbb{E}[Y_i^I] = \mathbb{E}[Y_i - Y_i^R] = \int_0^M (1 - G(z))\, \mathrm{d}z .$$

Example 1.4 Let S be the compound Poisson model with parameter λ and $\mathrm{Pa}(\alpha, \beta)$ distributed claim sizes. Assume that $\alpha > 1$, i.e. that $\mathbb{E}[Y_i] < \infty$. Let us compute the expected value of the outgo paid by the insurer;

$$\begin{aligned}\mathbb{E}[Y_i^I] &= \int_0^M \Bigl(\frac{\beta}{\beta + x}\Bigr)^\alpha \,\mathrm{d}x = \frac{\beta^\alpha}{\alpha - 1}\Bigl(\frac{1}{\beta^{\alpha-1}} - \frac{1}{(\beta + M)^{\alpha-1}}\Bigr) \\ &= \Bigl(1 - \Bigl(\frac{\beta}{\beta + M}\Bigr)^{\alpha-1}\Bigr)\frac{\beta}{\alpha - 1} = \Bigl(1 - \Bigl(\frac{\beta}{\beta + M}\Bigr)^{\alpha-1}\Bigr)\mathbb{E}[Y_i]\,.\end{aligned}$$

For $\mathbb{E}[S^I]$ it follows that

$$\mathbb{E}[S^I] = \Bigl(1 - \Bigl(\frac{\beta}{\beta + M}\Bigr)^{\alpha-1}\Bigr)\mathbb{E}[S]\,.$$

■

Let

$$N^R = \sum_{i=1}^N \mathbb{I}_{\{Y_i > M\}}$$

denote the number of claims the reinsurer has to pay for. We denote by $q = \mathbb{P}[Y_i > M]$ the probability that a claim amount exceeds the level M. What is the distribution of N^R? We first note that the moment generating function of $\mathbb{I}_{\{Y_i > M\}}$ is $q\mathrm{e}^r + 1 - q$.

(i) Let $N \sim \mathrm{B}(n, p)$. The moment generating function of N^R is

$$M_{N^R}(r) = (p(q\mathrm{e}^r + 1 - q) + 1 - p)^n = (pq\mathrm{e}^r + 1 - pq)^n\,.$$

Thus $N^R \sim \mathrm{B}(n, pq)$.

(ii) Let $N \sim \mathrm{Pois}(\lambda)$. The moment generating function of N^R is

$$M_{N^R}(r) = \exp\{\lambda((q\mathrm{e}^r + 1 - q) - 1)\} = \exp\{\lambda q(\mathrm{e}^r - 1)\}\,.$$

Thus $N^R \sim \mathrm{Pois}(\lambda q)$.

(iii) Let $N \sim \mathrm{NB}(\alpha, p)$. The moment generating function of N^R is

$$M_{N^R}(r) = \Bigl(\frac{p}{1 - (1 - p)(q\mathrm{e}^r + 1 - q)}\Bigr)^\alpha = \Bigl(\frac{\frac{p}{p+q-pq}}{1 - \Bigl(1 - \frac{p}{p+q-pq}\Bigr)\mathrm{e}^r}\Bigr)^\alpha.$$

Thus $N^R \sim \mathrm{NB}(\alpha, \frac{p}{p+q-pq})$.

1.8 Computation of the Distribution of S in the Discrete Case

Let us consider the case where the claim sizes $\{Y_i\}$ have an arithmetic distribution. We can assume that $\mathbb{P}[Y_i \in \mathbb{N}] = 1$ by choosing the monetary unit appropriately. We define $p_k = \mathbb{P}[N = k]$, $f_k = \mathbb{P}[Y_i = k]$ and $g_k = \mathbb{P}[S = k]$. For simplicity let us assume that $f_0 = 0$. Let $f_k^{*n} = \mathbb{P}[Y_1 + Y_2 + \cdots Y_n = k]$ denote the convolutions of the claim size distribution. Note that

$$f_k^{*(n+1)} = \sum_{i=1}^{k-1} f_i^{*n} f_{k-i} .$$

We get the following identities:

$$g_0 = \mathbb{P}[S = 0] = \mathbb{P}[N = 0] = p_0 ,$$

$$g_n = \mathbb{P}[S = n] = \mathbb{E}[\mathbb{P}[S = n \mid N]] = \sum_{k=1}^{n} p_k f_n^{*k} .$$

We have explicit formulae for the distribution of S, but the computation of the f_n^{*k}'s is messy. An easier procedure is called for. Let us now make an assumption on the distribution of N.

Assumption 1.5 Assume that there exist real numbers a and b such that

$$p_r = \Bigl(a + \frac{b}{r}\Bigr) p_{r-1}$$

for $r \in \mathbb{N} \setminus \{0\}$.

Let us check the assumption for the distributions of the models we have considered so far.

(i) **Binomial** $\mathrm{B}(n,p)$

$$\frac{p_r}{p_{r-1}} = \frac{\frac{n!}{r!(n-r)!} p^r (1-p)^{n-r}}{\frac{n!}{(r-1)!(n-r+1)!} p^{r-1} (1-p)^{n-r+1}} = \frac{(n-r+1)p}{r(1-p)}$$
$$= -\frac{p}{1-p} + \frac{(n+1)p}{r(1-p)} .$$

Thus

$$a = -\frac{p}{1-p} , \qquad b = \frac{(n+1)p}{1-p} .$$

(ii) **Poisson** Pois(λ)

$$\frac{p_r}{p_{r-1}} = \frac{\frac{\lambda^r}{r!}e^{-\lambda}}{\frac{\lambda^{r-1}}{(r-1)!}e^{-\lambda}} = \frac{\lambda}{r}\;.$$

Thus

$$a = 0\,, \qquad b = \lambda\,.$$

(iii) **Negative binomial** NB(α, p)

$$\frac{p_r}{p_{r-1}} = \frac{\frac{\Gamma(\alpha+r)}{r!\Gamma(\alpha)}p^\alpha(1-p)^r}{\frac{\Gamma(\alpha+r-1)}{(r-1)!\Gamma(\alpha)}p^\alpha(1-p)^{r-1}} = \frac{(\alpha+r-1)(1-p)}{r}$$
$$= 1-p+\frac{(\alpha-1)(1-p)}{r}\;.$$

Thus

$$a = 1-p\,, \qquad b = (\alpha-1)(1-p)\,.$$

These are in fact the only distributions satisfying the assumption. If we choose $a = 0$ we get by induction $p_r = p_0 b^r/r!$, which is the Poisson distribution.

If $a < 0$, then because $a + b/r$ is negative for r large enough there must be an $n_0 \in \mathbb{N}$ such that $b = -an_0$ in order that $p_r \geq 0$ for all r. Letting $n = n_0 - 1$ and $p = -a(1-a)$ we get the binomial distribution.

If $a > 0$ we need $a + b \geq 0$ in order that $p_1 \geq 0$. The case $a + b = 0$ can be considered as the degenerate case of a Poisson distribution with $\lambda = 0$. Suppose therefore that $a + b > 0$. In particular, $p_r > 0$ for all r. Let $k \in \mathbb{N} \setminus \{0\}$. Then

$$\sum_{r=1}^k rp_r = \sum_{r=1}^k (ar+b)p_{r-1} = a\sum_{r=1}^k (r-1)p_{r-1} + (a+b)\sum_{r=1}^k p_{r-1}$$
$$= a\sum_{r=1}^{k-1} rp_r + (a+b)\sum_{r=0}^{k-1} p_r\;.$$

This can be written as

$$kp_k = (a-1)\sum_{r=1}^{k-1} rp_r + (a+b)\sum_{r=0}^{k-1} p_r\;.$$

Suppose $a \geq 1$. Then $kp_k \geq (a+b)p_0$ and therefore $p_k \geq (a+b)p_0/k$. In particular,

$$1 - p_0 = \sum_{r=1}^\infty p_r \geq (a+b)p_0\sum_{r=1}^\infty \frac{1}{r}\;.$$

This is a contradiction. Thus $a < 1$. If we now let $p = 1 - a$ and $\alpha = a^{-1}(a + b)$ we get the negative binomial distribution.

We will use the following lemma.

Lemma 1.6 *Let $n \geq 2$. Then*

(i)
$$\mathbb{E}\Big[Y_1 \,\Big|\, \sum_{i=1}^{n} Y_i = r\Big] = \frac{r}{n}\,,$$

(ii)
$$p_n f_r^{*n} = \sum_{k=1}^{r-1}\Big(a + \frac{bk}{r}\Big) f_k p_{n-1} f_{r-k}^{*(n-1)}\,.$$

Proof (i) Noting that the $\{Y_i\}$ are iid. we get

$$\begin{aligned}\mathbb{E}\Big[Y_1 \,\Big|\, \sum_{i=1}^{n} Y_i = r\Big] &= \frac{1}{n}\sum_{j=1}^{n}\mathbb{E}\Big[Y_j \,\Big|\, \sum_{i=1}^{n} Y_i = r\Big]\\ &= \frac{1}{n}\mathbb{E}\Big[\sum_{j=1}^{n} Y_j \,\Big|\, \sum_{i=1}^{n} Y_i = r\Big] = \frac{r}{n}\,.\end{aligned}$$

(ii) Note that $f_0^{*(n-1)} = 0$. Thus

$$\begin{aligned}p_{n-1}\sum_{k=1}^{r-1}\Big(a + \frac{bk}{r}\Big) f_k f_{r-k}^{*(n-1)} &= p_{n-1}\sum_{k=1}^{r}\Big(a + \frac{bk}{r}\Big) f_k f_{r-k}^{*(n-1)}\\ &= p_{n-1}\sum_{k=1}^{r}\Big(a + \frac{bk}{r}\Big)\mathbb{P}\Big[Y_1 = k, \sum_{j=2}^{n} Y_j = r - k\Big]\\ &= p_{n-1}\sum_{k=1}^{r}\Big(a + \frac{bk}{r}\Big)\mathbb{P}\Big[Y_1 = k, \sum_{j=1}^{n} Y_j = r\Big]\\ &= p_{n-1}\sum_{k=1}^{r}\Big(a + \frac{bk}{r}\Big)\mathbb{P}\Big[Y_1 = k \,\Big|\, \sum_{j=1}^{n} Y_j = r\Big] f_r^{*n}\\ &= p_{n-1}\mathbb{E}\Big[a + \frac{bY_1}{r} \,\Big|\, \sum_{j=1}^{n} Y_j = r\Big] f_r^{*n}\\ &= p_{n-1}\Big(a + \frac{b}{n}\Big) f_r^{*n} = p_n f_r^{*n}\,.\end{aligned}$$

□

We now use the second formula to find a recursive expression for g_r. We know already that $g_0 = p_0$.

$$\begin{aligned}
g_r &= \sum_{n=1}^{\infty} p_n f_r^{*n} = p_1 f_r + \sum_{n=2}^{\infty} p_n f_r^{*n} \\
&= (a+b) p_0 f_r + \sum_{n=2}^{\infty} \sum_{i=1}^{r-1} \Bigl(a + \frac{bi}{r}\Bigr) f_i p_{n-1} f_{r-i}^{*(n-1)} \\
&= (a+b) g_0 f_r + \sum_{i=1}^{r-1} \Bigl(a + \frac{bi}{r}\Bigr) f_i \sum_{n=2}^{\infty} p_{n-1} f_{r-i}^{*(n-1)} \\
&= (a+b) f_r g_0 + \sum_{i=1}^{r-1} \Bigl(a + \frac{bi}{r}\Bigr) f_i g_{r-i} = \sum_{i=1}^{r} \Bigl(a + \frac{bi}{r}\Bigr) f_i g_{r-i} \,.
\end{aligned}$$

This is called the *Panjer recursion*. Note that the convolutions have vanished. Thus a simple recursion can be used to compute g_r.

Consider now the case where the claim size distribution is not arithmetic. Choose a bandwidth $\rho > 0$. Let

$$\bar{Y}_i = \inf\{n \in \mathbb{N} : Y_i \le \rho n\}\,, \quad \bar{S} = \sum_{i=1}^{N} \bar{Y}_i$$

and

$$\underline{Y_i} = \sup\{n \in \mathbb{N} : Y_i \ge \rho n\}\,, \quad \underline{S} = \sum_{i=1}^{N} \underline{Y_i}\,.$$

It is clear that

$$\rho \underline{S} \le S \le \rho \bar{S}\,.$$

The Panjer recursion can be used to compute the distribution of $\bar{S}$ and $\underline{S}$. Note that for $\underline{S}$ there is a positive probability of a claim of size 0. Thus the original distribution of N has to be changed slightly. The corresponding new claim number distribution can be found in Sect. 1.7.2.

1.9 Approximations to S

We have seen that the distribution of S is rather hard to determine, especially if no computers are available. Therefore approximations are called for.

1.9.1 The Normal Approximation

The number of claims is quite often large. In this case, for a deterministic number of claims, we would use a normal approximation. So we try a normal approximation to S.

We fit a normal distribution Z such that the first two moments coincide. Thus

$$\mathbb{P}[S \le x] \approx \mathbb{P}[Z \le x] = \Phi\Bigl(\frac{x - \mathbb{E}[N]\mu}{\sqrt{\operatorname{Var}[N]\mu^2 + \mathbb{E}[N](\mu_2 - \mu^2)}}\Bigr).$$

In particular, for a compound Poisson distribution the formula is

$$\mathbb{P}[S \le x] \approx \Phi\Bigl(\frac{x - \lambda\mu}{\sqrt{\lambda\mu_2}}\Bigr).$$

Example 1.7 Let S have a compound Poisson distribution with parameter $\lambda = 20$ and Pa(4,3) distributed claim sizes. Note that $\mu = 1$ and $\mu_2 = 3$.
Figure. 1.1 shows the densities of S (solid line) and of its normal approximation (dashed line).

Assume that we want to find a premium p such that $\mathbb{P}[S > p] \le 0.05$. Using the normal approximation we find that

$$\mathbb{P}[S > p] \approx 1 - \Phi\Bigl(\frac{p - 20}{\sqrt{60}}\Bigr) = 0.05\,.$$

Thus

$$\frac{p - 20}{\sqrt{60}} = 1.6449$$

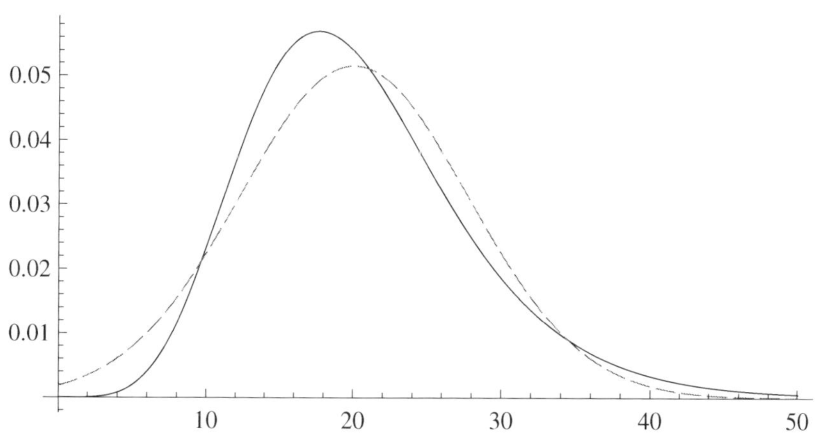

Fig. 1.1 Densities of S and its normal approximation

or $p = 32.7413$. If the accumulated claims should only exceed the premium in 1% of the cases we get

$$\frac{p-20}{\sqrt{60}} = 2.3263$$

or $p = 38.0194$. The exact values are 33.94 (5% quantile) and 42.99 (1% quantile). The approximation for the 5% quantile is quite good, but the approximation for the 1% quantile is not satisfactory. ∎

Problem: It often turns out that loss distributions are skewed. The normal distribution is not skewed. Hence we cannot hope that the approximation will be a good fit. We need an additional parameter to fit to the distribution.

1.9.2 The Translated Gamma Approximation

The idea of the translated gamma approximation is to approximate S by $k+Z$ where k is a real constant and Z has a gamma distribution $\Gamma(\gamma,\alpha)$. The first three moments should coincide. Note that $\mathbb{E}[((k+Z)-\mathbb{E}[k+Z])^3] = 2\gamma\alpha^{-3}$. Denote by $m = \mathbb{E}[S]$ the expected value of the risk, by σ^2 its variance and by β its coefficient of skewness. Then

$$\begin{aligned} k + \frac{\gamma}{\alpha} &= m\,, \\ \frac{\gamma}{\alpha^2} &= \sigma^2 \quad \text{and} \\ \frac{2\gamma\alpha^3}{\alpha^3\gamma^{3/2}} = \frac{2}{\sqrt{\gamma}} &= \beta\,. \end{aligned}$$

Solving this system yields

$$\gamma = \frac{4}{\beta^2}, \qquad \alpha = \frac{\sqrt{\gamma}}{\sigma} = \frac{2}{\beta\sigma}, \qquad k = m - \frac{\gamma}{\alpha} = m - \frac{2\sigma}{\beta}\,.$$

Remarks 1.8

(i) Note that β must be positive.
(ii) It may happen that k becomes negative. In that case the approximation has (like the normal approximation) some strictly positive mass on the negative half axis.
(iii) The gamma distribution is not easy to calculate. However if $2\gamma \in \mathbb{N}$ then $2\alpha Z \sim \chi^2_{2\gamma}$. The latter distribution can be found in statistical tables.

∎

Let us consider the compound Poisson case. Then

$$\gamma = \frac{4\lambda\mu_2^3}{\mu_3^2}, \quad \alpha = \frac{2\sqrt{\lambda\mu_2^3}}{\mu_3\sqrt{\lambda\mu_2}} = \frac{2\mu_2}{\mu_3}, \quad k = \lambda\mu - \frac{2\sqrt{\lambda\mu_2}\sqrt{\lambda\mu_2^3}}{\mu_3} = \lambda\Bigl(\mu - \frac{2\mu_2^2}{\mu_3}\Bigr).$$

Example 1.9 Again let S have a compound Poisson distribution with parameter $\lambda = 20$ and Pa(4,3) distributed claim sizes. A simple calculation shows that $\mu_3 = 27$. Thus

$$\gamma = \frac{80 \cdot 27}{729} = 2.96296, \quad \alpha = \frac{2 \cdot 3}{27} = 0.222222, \quad k = 20\Bigl(1 - \frac{2 \cdot 9}{27}\Bigr) = 6.66667.$$

Figure 1.2 shows the densities of S (solid line) and of its translated gamma approximation (dashed line).

Now $2\gamma = 5.92592 \approx 6$. Let $\bar{Z} \sim \chi_6^2$. Then

$$\mathbb{P}[S > p] \approx \mathbb{P}[Z > p - 6.66667] \approx \mathbb{P}[\bar{Z} > 0.444444(p - 6.66667)] = 0.05$$

and thus $0.444444(p - 6.66667) = 12.59$ or $p = 34.9942$. For the 1% case we obtain $0.444444(p - 6.66667) = 16.81$ or $p = 44.4892$. Because the coefficient of skewness $\beta = 1.1619$ is not small it is not surprising that the premium becomes larger than the normal approximation premium, especially if we are in the tail of the distribution. In any case the approximations of the quantiles are better than for the normal approximation. Recall that the exact values are 33.94 and 42.99 respectively. The approximation puts more weight in the tail of the distribution. In the domain where we did our computations the true value was overestimated. Note that for

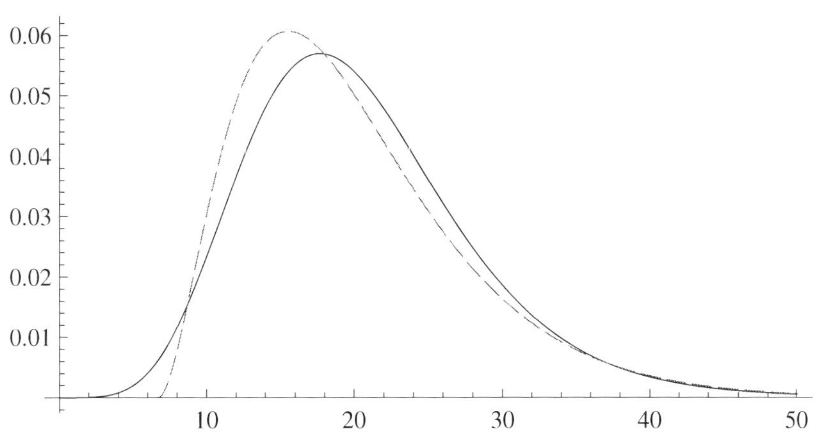

Fig. 1.2 Densities of S and its translated gamma approximation

smaller quantiles the true value will eventually be underestimated, because the tail of the correct distribution function tends to zero much more slowly than the tail of the gamma distribution. ■

1.9.3 The Edgeworth Approximation

Consider the random variable

$$Z = \frac{S - \mathbb{E}[S]}{\sqrt{\mathrm{Var}[S]}}\,.$$

The Taylor expansion of $\log M_Z(r)$ around $r = 0$ has the form

$$\log M_Z(r) = a_0 + a_1 r + a_2 \frac{r^2}{2} + a_3 \frac{r^3}{6} + a_4 \frac{r^4}{24} + \cdots$$

where

$$a_k = \frac{\mathrm{d}^k \log M_Z(r)}{\mathrm{d}r^k}\bigg|_{r=0}\,.$$

We already know that $a_0 = 0$, $a_1 = \mathbb{E}[Z] = 0$, $a_2 = \mathrm{Var}[Z] = 1$ and $a_3 = \mathbb{E}[(Z - \mathbb{E}[Z])^3] = \beta$. For a_4 a simple calculation shows that

$$\begin{aligned} a_4 &= \mathbb{E}[Z^4] - 4\mathbb{E}[Z^3]\mathbb{E}[Z] - 3\mathbb{E}[Z^2]^2 + 12\mathbb{E}[Z^2]\mathbb{E}[Z]^2 - 6\mathbb{E}[Z]^4 \\ &= \mathbb{E}[Z^4] - 3 = \frac{\mathbb{E}[(S - \mathbb{E}[S])^4]}{\mathrm{Var}[S]^2} - 3\,. \end{aligned}$$

We truncate the Taylor series after the term involving r^4. The moment generating function of Z can be written as

$$M_Z(r) \approx \mathrm{e}^{r^2/2}\mathrm{e}^{a_3 r^3/6 + a_4 r^4/24} \approx \mathrm{e}^{r^2/2}\Bigl(1 + a_3 \frac{r^3}{6} + a_4 \frac{r^4}{24} + a_3^2 \frac{r^6}{72}\Bigr)\,.$$

The inverse of $\exp\{r^2/2\}$ is easily found to be the normal distribution function $\Phi(x)$. For the other terms we derive

$$r^n \mathrm{e}^{r^2/2} = \int_{-\infty}^{\infty} \bigl(\mathrm{e}^{rx}\bigr)^{(n)} \Phi'(x)\,\mathrm{d}x = (-1)^n \int_{-\infty}^{\infty} \mathrm{e}^{rx} \Phi^{(n+1)}(x)\,\mathrm{d}x\,.$$

Thus the inverse of $r^n \mathrm{e}^{r^2/2}$ is $(-1)^n$ times the n-th derivative of Φ. The approximation yields

$$\mathbb{P}[Z \le z] \approx \Phi(z) - \frac{a_3}{6}\Phi^{(3)}(z) + \frac{a_4}{24}\Phi^{(4)}(z) + \frac{a_3^2}{72}\Phi^{(6)}(z)\,.$$

It can be computed that

$$\Phi^{(3)}(z) = \frac{1}{\sqrt{2\pi}}(z^2 - 1)e^{-z^2/2}\;,$$
$$\Phi^{(4)}(z) = \frac{1}{\sqrt{2\pi}}(-z^3 + 3z)e^{-z^2/2}\;,$$
$$\Phi^{(6)}(z) = \frac{1}{\sqrt{2\pi}}(-z^5 + 10z^3 - 15z)e^{-z^2/2}\;.$$

This approximation is called the *Edgeworth approximation*.

Consider now the compound Poisson case. One can see directly that $a_k = \lambda\mu_k(\lambda\mu_2)^{-k/2}$. Hence

$$\begin{aligned}
\mathbb{P}[S \le x] &= \mathbb{P}\Bigl[Z \le \frac{x-\lambda\mu}{\sqrt{\lambda\mu_2}}\Bigr] \\
&\approx \Phi\Bigl(\frac{x-\lambda\mu}{\sqrt{\lambda\mu_2}}\Bigr) - \frac{\mu_3}{6\sqrt{\lambda\mu_2^3}}\Phi^{(3)}\Bigl(\frac{x-\lambda\mu}{\sqrt{\lambda\mu_2}}\Bigr) + \frac{\mu_4}{24\lambda\mu_2^2}\Phi^{(4)}\Bigl(\frac{x-\lambda\mu}{\sqrt{\lambda\mu_2}}\Bigr) \\
&\quad + \frac{\mu_3^2}{72\lambda\mu_2^3}\Phi^{(6)}\Bigl(\frac{x-\lambda\mu}{\sqrt{\lambda\mu_2}}\Bigr)\;.
\end{aligned}$$

Example 1.10 We cannot consider Example 1.7 because $\mu_4 = \infty$. Let us therefore consider the compound Poisson distribution with $\lambda = 20$ and Pa(5,4) distributed claim sizes. Then we find $\mu = 1$, $\mu_2 = 8/3$, $\mu_3 = 16$ and $\mu_4 = 256$. Figure 1.3 gives us the densities of the exact distribution (solid line) and of its Edgeworth approximation (dashed line). The shape is approximated quite well close to the mean value 20. The tail is overestimated a little bit. Let us compare the 5 and 1% quantiles. Numerically we find 33.64 and 42.99, respectively. The exact values are 33.77 and 41.63, respectively. We see that the quantiles are approximated quite well. It is not

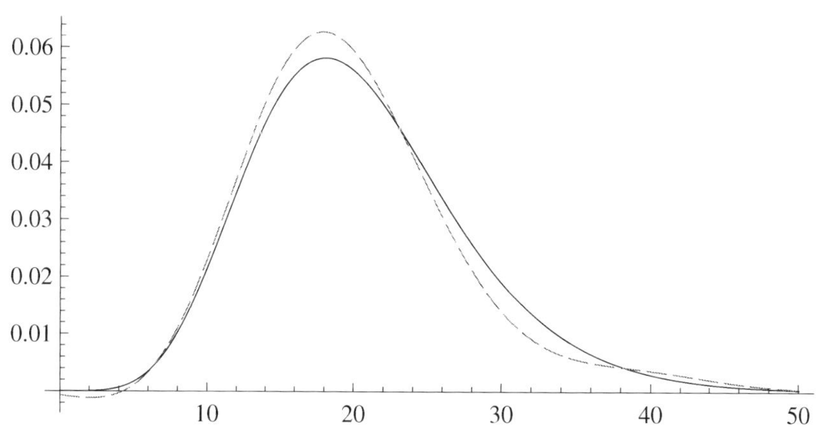

Fig. 1.3 Densities of S and its Edgeworth approximation

surprising that the 5% quantile is approximated better. The tail of the approximation decays like $x^5 e^{-cx^2}$ for some constant c while the exact tail decays like x^{-5}, which is much slower. ■

1.9.4 The Normal Power Approximation

The idea of the normal power approximation is to approximate

$$Z = \frac{S - \mathbb{E}[S]}{\sqrt{\mathrm{Var}[S]}} \approx \tilde{Z} + \frac{\beta_S}{6}(\tilde{Z}^2 - 1) ,$$

where $\tilde{Z}$ is a standard normal variable and β_S is the coefficient of skewness of S. The mean value of the approximation is 0, but the variance becomes $1 + \beta_S^2/18$ and the skewness is $\beta_S + \beta_S^3/27$. If β_S is not too large the approximation meets the first three moments of Z quite well. Let us now consider the inequality

$$\tilde{Z} + \frac{\beta_S}{6}(\tilde{Z}^2 - 1) \le z .$$

This is equivalent to

$$\Bigl(\tilde{Z} + \frac{3}{\beta_S}\Bigr)^2 \le 1 + \frac{6z}{\beta_S} + \frac{9}{\beta_S^2} .$$

We next consider positive $\tilde{Z} + 3/\beta_S$ only, because β_S should not be too large. Thus

$$\tilde{Z} + \frac{3}{\beta_S} \le \sqrt{1 + \frac{6z}{\beta_S} + \frac{9}{\beta_S^2}} .$$

We therefore get the approximation

$$\mathbb{P}[S \le x] \approx \Phi\biggl(\sqrt{1 + \frac{6(x - \mathbb{E}[S])}{\beta_S\sqrt{\mathrm{Var}[S]}} + \frac{9}{\beta_S^2}} - \frac{3}{\beta_S}\biggr) .$$

In the compound Poisson case the approximation reads

$$\mathbb{P}[S \le x] \approx \Phi\biggl(\sqrt{1 + \frac{6(x - \lambda\mu)\mu_2}{\mu_3} + \frac{9\lambda\mu_2^3}{\mu_3^2}} - \frac{3\sqrt{\lambda\mu_2^3}}{\mu_3}\biggr) .$$

Example 1.11 Again let S have a compound Poisson distribution with parameter $\lambda = 20$ and Pa(4,3) distributed claim sizes. Plugging in the parameters we find the approximation

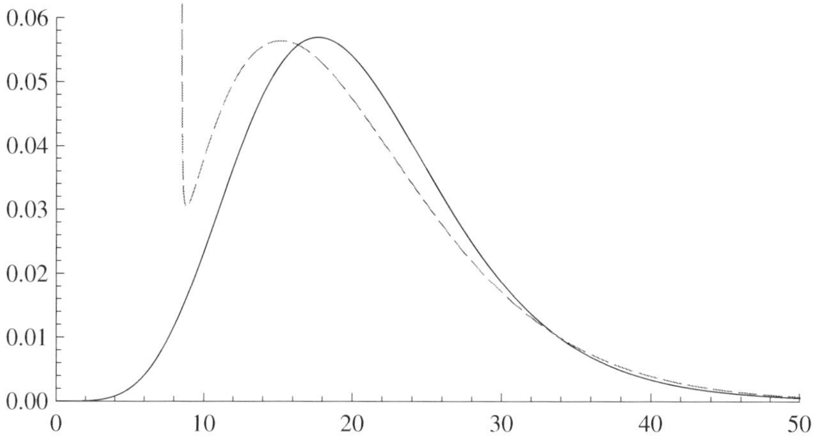

Fig. 1.4 Densities of S and its normal power approximation

$$\begin{aligned}\mathbb{P}[S \le x] &\approx \Phi\left(\sqrt{1 + \frac{6(x-20)3}{27} + \frac{9 \cdot 20 \cdot 3^3}{27^2}} - \frac{3\sqrt{20 \cdot 3^3}}{27}\right) \\ &= \Phi\left(\sqrt{\frac{2x-17}{3}} - \sqrt{\frac{20}{3}}\right) .\end{aligned}$$

The approximation is only valid if $x \ge 8.5$. The density of S (solid line) and of the normal power approximation (dashed line) are given in Fig. 1.4. The density of the approximation has a pole at $x = 8.5$. We calculate the 5% quantile, which is 35.2999. The 1% quantile is 44.6369. The approximation overestimates the true values by about 4%. As we can also see by comparison with Fig. 1.2, the translated gamma approximation fits the tail much better than the normal power approximation. ■

1.10 Premium Calculation Principles

Let the annual premium for a certain risk be π, and denote the losses in year i by S_i and assume that they are iid. The company has a certain initial capital w. Then the capital of the company after year i is

$$X_i = w + \pi i - \sum_{j=1}^{i} S_j .$$

The stochastic process $\{X_i\}$ is called a random walk. From Lemma E.1 we conclude that

(i) if $\pi < \mathbb{E}[S_1]$ then X_i converges to $-\infty$ a.s.,
(ii) if $\pi = \mathbb{E}[S_1]$ then a.s.

$$\limsup_{i\to\infty} X_i = -\liminf_{i\to\infty} X_i = \infty ,$$

(iii) if $\pi > \mathbb{E}[S_1]$ then X_i converges to ∞ a.s. and there is a strictly positive probability that $X_i \geq 0$ for all $i \in \mathbb{N}$.

As an insurer we only have a chance to survive ultimately if $\pi > \mathbb{E}[S]$. Therefore the latter must hold for any reasonable premium calculation principle. Another argument will be given in Chap. 2.

1.10.1 The Expected Value Principle

The most popular premium calculation principle is

$$\pi = (1+\theta)\mathbb{E}[S] ,$$

where θ is a strictly positive parameter called the *safety loading*. The premium is very easy to calculate. One only has to estimate the quantity $\mathbb{E}[S]$. A disadvantage is that the principle is not sensitive to heavy tailed distributions.

1.10.2 The Variance Principle

A way of making the premium principle more sensitive to higher risks is to use the variance for determining the risk loading, i.e.

$$\pi = \mathbb{E}[S] + \alpha \mathrm{Var}[S]$$

for a strictly positive parameter α.

1.10.3 The Standard Deviation Principle

The security loading in the variance principle is not stable with respect to the monetary unit. A principle similar to the variance principle without this problem is

$$\pi = \mathbb{E}[S] + \beta\sqrt{\mathrm{Var}[S]}$$

for a strictly positive parameter β. Observe that the loss can be written as

$$S - \pi = \sqrt{\mathrm{Var}[S]}\left(\frac{S - \mathbb{E}[S]}{\sqrt{\mathrm{Var}[S]}} - \beta\right).$$

The standardized loss is therefore a random variable with mean 0 and variance 1 minus the risk loading parameter.

1.10.4 The Modified Variance Principle

Another possibility is to change the variance principle to

$$\pi = \mathbb{E}[S] + \alpha\frac{\mathrm{Var}[S]}{\mathbb{E}[S]}$$

for some $\alpha > 0$.

1.10.5 The Principle of Zero Utility

This premium principle will be discussed in detail in Chap. 2. Denote by w the initial capital of the insurance company. The worst thing that may happen for the company is a very high accumulated sum of claims. Therefore high losses should be given a stronger weighting than small losses. Hence the company chooses a utility function v, which should have the following properties:

- $v(0) = 0$,
- $v(x)$ is strictly increasing,
- $v(x)$ is strictly concave.

The first property is for convenience only, the second one means that smaller losses are preferred to higher losses, and the last condition gives stronger weights to higher losses.

The premium for the next year is computed as the value π such that

$$v(w) = \mathbb{E}[v(w + \pi - S)]. \tag{1.3}$$

This means, the expected utility is the same whether the insurance contract is taken or not.

Lemma 1.12

(i) If a solution to (1.3) *exists, then it is unique.*
(ii) If $\mathbb{P}[S = \mathbb{E}[S]] < 1$ *then* $\pi > \mathbb{E}[S]$.
(iii) The premium is independent of w *for all loss distributions if and only if* $v(x) = A(1 - \mathrm{e}^{-\alpha x})$ *for some* $A > 0$ *and* $\alpha > 0$.

Proof (i) Let $\pi_1 > \pi_2$ be two solutions to (1.3). Then because $v(x)$ is strictly increasing

$$v(w) = \mathbb{E}[v(w + \pi_1 - S)] > \mathbb{E}[v(w + \pi_2 - S)] = v(w) ,$$

which is a contradiction.

(ii) By Jensen's inequality

$$v(w) = \mathbb{E}[v(w + \pi - S)] < v(w + \pi - \mathbb{E}[S])$$

and thus because $v(x)$ is strictly increasing we get $w + \pi - \mathbb{E}[S] > w$.

(iii) A simple calculation shows that the premium is independent of w if $v(x) = A(1 - \mathrm{e}^{-\alpha x})$. Assume now that the premium is independent of w. Let $\mathbb{P}[S = 1] = 1 - \mathbb{P}[S = 0] = q$ and denote by $\pi(q)$ the corresponding premium. Then

$$qv(w + \pi(q) - 1) + (1 - q)v(w + \pi(q)) = v(w) . \tag{1.4}$$

Note that as a concave function $v(x)$ is differentiable almost everywhere. Taking the derivative of (1.4) (at points where it is allowed) with respect to w yields

$$qv'(w + \pi(q) - 1) + (1 - q)v'(w + \pi(q)) = v'(w) . \tag{1.5}$$

Varying q, the function $\pi(q)$ takes all values in $(0, 1)$. Changing w and q in such a way that $w + \pi(q)$ remains constant shows that $v'(w)$ is continuous. Thus $v(w)$ must be continuously differentiable everywhere.

The derivative of (1.4) with respect to q is

$$v(w+\pi(q)-1)-v(w+\pi(q))+\pi'(q)[qv'(w+\pi(q)-1)+(1-q)v'(w+\pi(q))] = 0 . \tag{1.6}$$

The implicit function theorem gives that $\pi(q)$ is indeed differentiable. Plugging (1.5) into (1.6) yields

$$v(w + \pi(q) - 1) - v(w + \pi(q)) + \pi'(q)v'(w) = 0 . \tag{1.7}$$

Note that $\pi'(q) > 0$ follows immediately. The derivative of (1.7) with respect to w is

$$v'(w + \pi(q) - 1) - v'(w + \pi(q)) + \pi'(q)v''(w) = 0 ,$$

showing that $v(w)$ is twice continuously differentiable. From the derivative of (1.7) with respect to q it follows that

$$\begin{aligned}\pi'(q)[v'(w+\pi(q)-1)-v'(w+\pi(q))]+\pi''(q)v'(w)\\ =-\pi'(q)^2v''(w)+\pi''(q)v'(w)=0\,.\end{aligned}$$

Thus $\pi''(q)\le 0$ and

$$\frac{v''(w)}{v'(w)}=\frac{\pi''(q)}{\pi'(q)^2}\,.$$

The left-hand side is independent of q and the right-hand side is independent of w. Thus it must be constant. Hence

$$\frac{v''(w)}{v'(w)}=-\alpha$$

for some $\alpha\ge 0$. $\alpha=0$ would not lead to a strictly concave function. Solving the differential equation proves the assertion. □

1.10.6 The Mean Value Principle

In contrast to the principle of zero utility it would be desirable to have a premium principle which is independent of the initial capital. The insurance company values its losses and compares it to the loss of the customer who has to pay its premium. Because high losses are less desirable they need to have a higher value. Similarly to the utility function we choose a value function v with properties

- $v(0)=0$,
- $v(x)$ is strictly increasing and
- $v(x)$ is strictly convex.

The premium is determined by the equation

$$v(\pi)=\mathbb{E}[v(S)]$$

or equivalently

$$\pi=v^{-1}(\mathbb{E}[v(S)])\,.$$

Again, it follows from Jensen's inequality that $\pi>\mathbb{E}[S]$ provided $\mathbb{P}[S=\mathbb{E}[S]]<1$.

1.10.7 The Exponential Principle

Let $\alpha > 0$ be a parameter. Choose the utility function $v(x) = 1 - \mathrm{e}^{-\alpha x}$. The premium can be obtained

$$1 - \mathrm{e}^{-\alpha w} = \mathbb{E}\left[1 - \mathrm{e}^{-\alpha(w+p-S)}\right] = 1 - \mathrm{e}^{-\alpha w}\mathrm{e}^{-\alpha p}\mathbb{E}\left[\mathrm{e}^{\alpha S}\right]$$

from which it follows that

$$\pi = \frac{\log(M_S(\alpha))}{\alpha}. \tag{1.8}$$

The same premium formula can be obtained by using the mean value principle with value function $v(x) = \mathrm{e}^{\alpha x} - 1$. Note that the principle can only be applied if the moment generating function of S exists at the point α.

How does the premium depend on the parameter α? In order to answer this question the following lemma is proved.

Lemma 1.13 *For a random variable X with moment generating function $M_X(r)$ the function* $\log M_X(r)$ *is convex. If X is not deterministic then* $\log M_X(r)$ *is strictly convex.*

Proof Let $F(x)$ denote the distribution function of X. The second derivative of $\log M_X(r)$ is

$$(\log M_X(r))'' = \frac{M_X''(r)}{M_X(r)} - \left(\frac{M_X'(r)}{M_X(r)}\right)^2.$$

Consider

$$\frac{M_X''(r)}{M_X(r)} = \frac{\frac{\mathrm{d}^2}{\mathrm{d}r^2}\int_{-\infty}^{\infty} \mathrm{e}^{rx}\,\mathrm{d}F(x)}{M_X(r)} = \frac{\int_{-\infty}^{\infty} x^2\mathrm{e}^{rx}\,\mathrm{d}F(x)}{M_X(r)}$$

and

$$\frac{M_X'(r)}{M_X(r)} = \frac{\int_{-\infty}^{\infty} x\mathrm{e}^{rx}\,\mathrm{d}F(x)}{M_X(r)}.$$

Because

$$\frac{\int_{-\infty}^{\infty} \mathrm{e}^{rx}\,\mathrm{d}F(x)}{M_X(r)} = \frac{M_X(r)}{M_X(r)} = 1$$

it follows that $F_r(x) = (M_X(r))^{-1}\int_{-\infty}^{x} \mathrm{e}^{ry}\,\mathrm{d}F(y)$ is a distribution function. Let Z be a random variable with distribution function F_r. Then

$$(\log M_X(r))'' = \mathrm{Var}[Z] \geq 0.$$

The variance is strictly larger than 0 if Z is not deterministic and the latter is equivalent to X not being deterministic. □

The question of how the premium depends on α can now be answered.

Lemma 1.14 *Let $\pi(\alpha)$ be the premium determined by the exponential principle. Assume that $M_S''(\alpha) < \infty$. The function $\pi(\alpha)$ is strictly increasing provided S is not deterministic.*

Proof Take the derivative of (1.8)

$$\pi'(\alpha) = \frac{M_S'(\alpha)}{\alpha M_S(\alpha)} - \frac{\log(M_S(\alpha))}{\alpha^2} = \frac{1}{\alpha}\Bigl(\frac{M_S'(\alpha)}{M_S(\alpha)} - \pi(\alpha)\Bigr).$$

Furthermore,

$$\begin{aligned}\frac{\mathrm{d}}{\mathrm{d}\alpha}(\alpha^2\pi'(\alpha)) &= \frac{M_S'(\alpha)}{M_S(\alpha)} - \pi(\alpha) + \alpha\Bigl(\frac{M_S''(\alpha)}{M_S(\alpha)} - \Bigl(\frac{M_S'(\alpha)}{M_S(\alpha)}\Bigr)^2\Bigr) - \alpha\pi'(\alpha)\\ &= \alpha\Bigl(\frac{M_S''(\alpha)}{M_S(\alpha)} - \Bigl(\frac{M_S'(\alpha)}{M_S(\alpha)}\Bigr)^2\Bigr).\end{aligned}$$

From Lemma 1.13 it follows that

$$\frac{\mathrm{d}}{\mathrm{d}\alpha}(\alpha^2\pi'(\alpha)) \gtreqless 0 \text{ iff } \alpha \gtreqless 0.$$

Thus the function $\alpha^2\pi'(\alpha)$ has a minimum at 0. Because its value at 0 vanishes it follows that $\pi'(\alpha) > 0$. $\pi'(0) = \frac{1}{2}\mathrm{Var}[S] > 0$ can be verified directly using L'Hospital's rule. □

1.10.8 The Esscher Principle

A simple idea for pricing is to use a distribution $\tilde{F}(x)$ which is related to the distribution function of the risk $F(x)$, but gives more weight to larger losses. If exponential moments exist, a natural choice would be to consider the exponential class

$$F_\alpha(x) = (M_S(\alpha))^{-1}\int_{-\infty}^{x} \mathrm{e}^{ry}\,\mathrm{d}F_S(y).$$

This yields the Esscher premium principle

$$\pi = \frac{\mathbb{E}[S\mathrm{e}^{\alpha S}]}{M_S(\alpha)},$$

where $\alpha > 0$. In order that the principle can be used we have to show that $\pi > \mathbb{E}[S]$ if $\mathrm{Var}[S] > 0$. Indeed, $\pi = (\log M_S(r))'|_{r=\alpha}$ and $\log M_S(r)$ is a strictly convex function by Lemma 1.13. Thus $(\log M_S(r))'$ is strictly increasing, in particular $\pi > (\log M_S(r))'|_{r=0} = \mathbb{E}[S]$.

1.10.9 The Distortion Principle

A disadvantage of the Esscher principle is that $M_S(\alpha)$ has to exist. We therefore look for a way to increase the weight for large claims, so that not necessarily exponential moments need to exist. One possibility is to change the distribution function $F_S(x)$ by a distortion function $g(x)$, that is, $F_g(x) = 1 - g(1 - F_S(x))$. Here, $g(x)$ is an increasing function with $g(0) = 0$ and $g(1) = 1$, such that $g(x) > x$ for $x \in (0, 1)$. This yields the premium

$$\pi = \int_0^\infty g(1 - F_S(x))\,\mathrm{d}x - \int_{-\infty}^0 [1 - g(1 - F_S(x))]\,\mathrm{d}x\,.$$

Note that $\pi > \mathbb{E}[S]$, provided $\mathrm{Var}[S] > 0$. Because $\mathrm{d}F_g(x) = g'(1 - F_S(x))\,\mathrm{d}F_S(x)$ it follows that the distributions $F_g(x)$ and $F_S(x)$ are equivalent if $g(x)$ is absolutely continuous.

Example 1.15 A simple example is the proportional hazard transform principle. The function g is chosen as $g(x) = x^r$ for some $r \in (0, 1)$. The premium becomes

$$\pi = \int_0^\infty (1 - F_S(x))^r\,\mathrm{d}x - \int_{-\infty}^0 1 - (1 - F_S(x))^r\,\mathrm{d}x\,.$$

A sufficient condition for π to be finite would be that $\mathbb{E}[\max\{S, 0\}^\alpha] < \infty$ for some $\alpha > 1/r$. ∎

1.10.10 The Percentage Principle

A company wants to keep the risk that the accumulated sum of claims exceeds the premium income small. They choose a parameter ε and the premium

$$\pi = \inf\{x > 0 : \mathbb{P}[S > x] \le \varepsilon\}\,.$$

It may be possible that $\pi < \mathbb{E}[S]$. If the claims are bounded it would be possible to choose $\varepsilon = 0$. This principle is called the *maximal loss principle*. It is clear that the latter cannot be of interest, even though it has been used for criminal purposes, and not only in medieval times.

1.10.11 Desirable Properties

- **Simplicity**: The premium calculation principle should be easy to calculate. This is the case for principles 1.10.1–1.10.4.

- **Consistency**: If a deterministic constant is added to the risk then the premium should increase by the same constant. This is the case for principles 1.10.2, 1.10.3, 1.10.5, 1.10.7–1.10.10. The property only holds for the mean value principle if it has an exponential value function (exponential principle) or a linear value function. This can be seen by a proof similar to that of Lemma 1.12 iii).
- **Additivity**: If a portfolio consists of two independent risks then its premium should be the sum of the individual premia. This is the case for principles 1.10.1, 1.10.2, 1.10.7 and 1.10.8. For Sects. 1.10.5 and 1.10.6 it is only the case if they coincide with the exponential principle or if $\pi = \mathbb{E}[S]$. A proof can be found in [65, p. 75].

1.11 Risk Measures

1.11.1 Introduction

In order to measure the risk, we have seen that utility theory yields a way to compare risks. The regulator has to determine a regulatory capital that a risk bearer needs to hold. A simple idea would be that the regulator determines a utility function. The regulatory capital would then be (minus) the capital that gives the same utility. However, not all utility functions lead to principles that are satisfactory for supervising purposes. We now want to consider the problem more generally.

We consider here a financial position, that is, a random variable X. In our case, X will be an annual profit of the insurer. Thus, the insurer has a loss if $X < 0$. We assume for simplicity of the presentation that $X \in \mathcal{L}^\infty$, that is, there exists a number $a_X \in \mathbb{R}$ such that $\mathbb{P}[|X| \le a_X] = 1$. A similar theory can be developed for $X \in \mathcal{L}^p$ for some $p \ge 1$, that is, the variables with $\mathbb{E}[|X|^p] < \infty$. We assign to each risk a regulatory capital, $\rho : \mathcal{L}^\infty \to \mathbb{R}$.

We define the following properties:

- ρ is called *monotone* if $X \le Y$ implies that $\rho(X) \ge \rho(Y)$. That is, a larger profit means less regulatory capital.
- ρ is called *normed* if $\rho(0) = 0$. No risk means no regulatory capital.
- ρ is called *translation invariant* if $\rho(X - a) = \rho(X) + a$ for all $a \in \mathbb{R}$. Thus, promising a deterministic capital increases the regulatory capital by the same amount. If we add the regulatory capital to the profit we will end with the same risk.
- ρ is called *positive homogeneous* if $\rho(\lambda X) = \lambda\rho(X)$ for all $\lambda \ge 0$. So, doubling a portfolio means double regulatory capital.
- ρ is called *subadditive* if $\rho(X + Y) \le \rho(X) + \rho(Y)$. This is something like a diversification effect. In particular, it means that it is not possible to split a risk in order to reduce the regulatory capital.
- ρ is called *convex* if $\rho(\lambda X + (1 - \lambda)Y) \le \lambda\rho(X) + (1 - \lambda)\rho(Y)$ for all $\lambda \in (0, 1)$ and all $X, Y \in \mathcal{L}^\infty$. This means that diversification leads to lower regulatory capital.

Definition 1.16 ρ is called a *monetary risk measure* if it is normed, monotone and translation invariant. A monetary risk measure ρ is called *coherent* if it is positive homogeneous and subadditive. A monetary risk measure ρ is called *convex* if it is convex.

Example 1.17 A popular example is the *Value-at-Risk* (V@R). The Value-at-Risk to the confidence level $\alpha \in (0, 1)$ is defined as

$$\mathrm{V@R}_\alpha(X) = \inf\{x \in \mathbb{R} : \mathbb{P}[x + X \geq 0] \geq \alpha\} .$$

In the Solvency II rules, α is chosen to be 0.995. This means that the regulatory capital should be insufficient only once in two hundred years. From the definition, it is clear that V@R is monotone and normed. It is easy to see that V@R is translation invariant. Thus, V@R is a monetary risk measure. Further, positive homogeneity also follows readily. However, V@R is neither subadditive nor convex. Consider the following simple example. Let X_k be independent with $\mathbb{P}[X_k = 1] = \kappa\alpha = 1 - \mathbb{P}[X_k = 0]$ for a constant $1 < \kappa < \alpha^{-\frac{1}{2}}$. Then $\mathrm{V@R}_\alpha(X_k) = -1$. Further, $\mathbb{P}[X_1 + X_2 = 2] = \kappa^2\alpha^2 < \alpha$, $\mathbb{P}[X_1 + X_2 = 1] = 2\kappa\alpha(1 - \kappa\alpha)$ and $\mathbb{P}[X_1 + X_2 = 0] = (1 - \kappa\alpha)^2$. Hence, $\mathrm{V@R}(X_1 + X_2) \geq -1 > -2 = \mathrm{V@R}(X_1) + \mathrm{V@R}(X_2)$. This shows that V@R is not subadditive. That ρ is not convex follows from the lemma below.

A similar monetary risk measure is the maximal loss $\mathrm{V@R}_1$. It follows readily that the maximal loss is coherent. ■

We next show that a coherent risk measure is convex.

Lemma 1.18 *A monetary risk measure ρ that is positive homogeneous is coherent if and only if it is convex.*

Proof Let ρ be coherent. Then

$$\rho(\lambda X + (1 - \lambda)Y) \leq \rho(\lambda X) + \rho((1 - \lambda)Y) = \lambda\rho(X) + (1 - \lambda)\rho(Y) ,$$

and ρ is convex. If ρ is convex, then

$$\rho(X + Y) = 2\rho\Bigl(\frac{1}{2}X + \frac{1}{2}Y\Bigr) \leq 2\Bigl(\frac{1}{2}\rho(X) + \frac{1}{2}\rho(Y)\Bigr) = \rho(X) + \rho(Y) .$$

Thus, ρ is subadditive and therefore coherent. □

Example 1.19 We use the exponential utility principle with parameter $\beta > 0$. Define $\rho(X) = \beta^{-1} \log \mathbb{E}[e^{-\beta X}]$. Then ρ is monotone and normed. Further

$$\begin{aligned}\rho(X - a) &= \beta^{-1} \log \mathbb{E}[e^{-\beta(X-a)}] = \beta^{-1} \log e^{\beta a}\mathbb{E}[e^{-\beta X}] \\ &= a + \beta^{-1} \log \mathbb{E}[e^{-\beta X}] = \rho(X) + a .\end{aligned}$$

This shows that ρ is a monetary risk measure. Define

$$f(\lambda) = \rho(\lambda X + (1-\lambda)Y) = \beta^{-1} \log \mathbb{E}[e^{-\beta(\lambda X+(1-\lambda)Y)}] .$$

We find

$$f'(\lambda) = -\frac{\mathbb{E}[(X-Y)e^{-\beta(\lambda X+(1-\lambda)Y)}]}{\mathbb{E}[e^{-\beta(\lambda X+(1-\lambda)Y)}]} ,$$

and the second derivative

$$\beta\Big\{\frac{\mathbb{E}[(X-Y)^2 e^{-\beta(\lambda X+(1-\lambda)Y)}]}{\mathbb{E}[e^{-\beta(\lambda X+(1-\lambda)Y)}]} - \Big(\frac{\mathbb{E}[(X-Y)e^{-\beta(\lambda X+(1-\lambda)Y)}]}{\mathbb{E}[e^{-\beta(\lambda X+(1-\lambda)Y)}]}\Big)^2\Big\} .$$

The expression between the brackets is the variance of $(X-Y)$ under the absolutely continuous measure

$$\frac{\mathrm{d}Q}{\mathrm{d}\mathbb{P}} = \frac{e^{-\beta(\lambda X+(1-\lambda)Y)}}{\mathbb{E}[e^{-\beta(\lambda X+(1-\lambda)Y)}]} ,$$

and therefore positive. Thus, $f(\lambda)$ is a convex function. If $X-Y$ is not deterministic, then $f(\lambda)$ is strictly convex. In particular,

$$\rho(\lambda X + (1-\lambda)Y) = f(\lambda) \le \lambda f(1) + (1-\lambda)f(0) = \lambda\rho(X) + (1-\lambda)\rho(Y) .$$

This shows that the measure is convex. Choosing $Y = 0$ and X not deterministic, then by the strict convexity

$$\rho(\lambda X) = f(\lambda) < \lambda f(1) + (1-\lambda)f(0) = \lambda\rho(X) + (1-\lambda)\rho(0) = \lambda\rho(X) .$$

This shows that ρ is not positive homogeneous, and therefore not coherent. ■

There are many applications of risk measures in insurance such as, for example, premium principles or allocation of economic capital. More on this topic can be found, for example, in [45].

1.11.2 Representation of Convex and Coherent Risk Measures

Let $\mathcal{M}$ be the set of probability measures that are absolutely continuous with respect to $\mathbb{P}$. By $\mathbb{E}_Q$ we denote expectation with respect to the measure Q, $\mathbb{E}_Q[X] = \mathbb{E}[X\frac{\mathrm{d}Q}{\mathrm{d}\mathbb{P}}]$.

Theorem 1.20 *A mapping $\rho : \mathcal{L}^\infty \to \mathbb{R}$ is a convex risk measure if and only if it is of the form*

$$\rho(X) = -\inf_{Q\in\mathcal{P}} \{\mathbb{E}_Q[X] + \alpha(Q)\} ,$$

where $\emptyset \ne \mathcal{P} \subset \mathcal{M}$ and $\alpha : \mathcal{P} \to \mathbb{R}$ is a mapping such that $\inf_{Q\in\mathcal{P}} \alpha(Q) = 0$.

Proof We prove here only the easy direction. The complete proof can, for example, be found in [61]. Suppose that ρ has the proposed representation. That the mapping is monotone and translation invariant is easy to see. By the assumption on α, we have

$$\rho(0) = -\inf_{Q\in\mathcal{P}}\{\mathbb{E}_Q[0] + \alpha(Q)\} = -\inf_{Q\in\mathcal{P}}\alpha(Q) = 0\,.$$

Thus ρ is a monetary risk measure. Choose now $X, Y \in \mathcal{L}^\infty$ and $\lambda \in (0, 1)$. Let $Z = \lambda X + (1-\lambda)Y$. Let $\varepsilon > 0$. There is a $Q \in \mathcal{P}$ such that $\rho(Z) < -\mathbb{E}_Q[Z] - \alpha(Q) + \varepsilon$. Then

$$\rho(Z) - \varepsilon < -\lambda(\mathbb{E}_Q[X] + \alpha(Q)) - (1-\lambda)(\mathbb{E}_Q[Y] + \alpha(Q)) \le \lambda\rho(X) + (1-\lambda)\rho(Y)\,.$$

Since ε is arbitrary, we have $\rho(Z) \le \lambda\rho(X) + (1-\lambda)\rho(Y)$. Thus ρ is convex. □

Corollary 1.21 *A mapping $\rho : \mathcal{L}^\infty \to \mathbb{R}$ is a coherent risk measure if and only if it is of the form*

$$\rho(X) = -\inf_{Q\in\mathcal{P}}\mathbb{E}_Q[X]\,,$$

where $\emptyset \ne \mathcal{P} \subset \mathcal{M}$.

Proof If the representation holds, ρ is a convex risk measure. It is easy to see that it is also positive homogeneous, and therefore a coherent risk measure. Suppose that ρ is coherent and therefore also convex. Then there is a set $\mathcal{P} \subset \mathcal{M}$ and a mapping $\alpha : \mathcal{P} \to \mathbb{R}$ such that the representation for convex risk measures holds. Note that $\alpha(Q) \ge 0$. We have

$$\begin{aligned}\rho(X) &= \lambda^{-1}\rho(\lambda X) = -\lambda^{-1}\inf_{Q\in\mathcal{P}}\{\mathbb{E}_Q[\lambda X] + \alpha(Q)\} \\ &= -\inf_{Q\in\mathcal{P}}\{\mathbb{E}_Q[X] + \lambda^{-1}\alpha(Q)\} = -\inf_{\lambda\ge 0}\inf_{Q\in\mathcal{P}}\{\mathbb{E}_Q[X] + \lambda^{-1}\alpha(Q)\} \\ &= -\inf_{Q\in\mathcal{P}}\inf_{\lambda\ge 0}\{\mathbb{E}_Q[X] + \lambda^{-1}\alpha(Q)\} = -\inf_{Q\in\mathcal{P}}\mathbb{E}_Q[X]\,.\end{aligned}$$

□

Example 1.22 We define the *Average-Value-at-Risk*

$$\mathrm{AV@R}_\alpha(X) = \frac{1}{1-\alpha}\int_\alpha^1 \mathrm{V@R}_\beta(X)\,d\beta\,,$$

for an $\alpha \in (0, 1)$. For the Swiss Solvency Test, α is chosen to be 0.99. From the corresponding properties of the Value-at-Risk we conclude that $\mathrm{AV@R}_\alpha$ is a positive homogeneous monetary risk measure. Denote by F the distribution function of $-X$. Then $\mathrm{V@R}_\beta(X) = F^{-1}(\beta)$. Let U be uniformly distributed on $(0, 1)$. Note that $F^{-1}(U)$ has the same distribution as $-X$. Thus

$$\begin{aligned}\mathrm{AV@R}_\alpha(X) &= \frac{1}{1-\alpha}\int_\alpha^1 F^{-1}(\beta)\,\mathrm{d}\beta = \frac{1}{1-\alpha}\int_0^1 F^{-1}(\beta)1\!\!1_{\beta>\alpha}\,\mathrm{d}\beta \\ &= \frac{1}{1-\alpha}\mathbb{E}[F^{-1}(U)1\!\!1_{U>\alpha}]\,.\end{aligned}$$

The variable $1\!\!1_{U>\alpha}$ is measurable with respect to $\sigma(F^{-1}(U))$. One can therefore show that $F^{-1}(U)$ and $1\!\!1_{U>\alpha}$ are maximally correlated under all variables Y and Z such that Y has distribution function F and Z is Bernoulli-distributed with $\mathbb{P}[Z=0] = 1-\mathbb{P}[Z=1] = \alpha$. Hence

$$\mathrm{AV@R}_\alpha(X) = \sup_Z \mathbb{E}\Big[(-X)\frac{Z}{1-\alpha}\Big]\,,$$

where Z is Bernoulli-distributed with $\mathbb{E}[Z] = 1-\alpha$. This yields

$$\mathcal{P} = \{Q : \mathrm{d}Q = Z/(1-\alpha)\,\mathrm{d}\mathbb{P},\ Z \text{ is Bernoulli } (1-\alpha) \text{ distributed}\}\,.$$

Therefore

$$\mathrm{AV@R}_\alpha(X) = \sup_{Q\in\mathcal{P}} \mathbb{E}_Q[-X] = -\inf_{Q\in\mathcal{P}} \mathbb{E}_Q[X]\,,$$

which gives that $\mathrm{AV@R}_\alpha$ is coherent. ■

Bibliographical Remarks

For further literature on the individual model (Sect. 1.6) see also [65, p. 51]. A reference to why reinsurance is necessary is [118]. Further literature on reinsurance may be found, for example, in [32, 47, 53, 70, 78, 88]. The Panjer recursion was originally described in [106], see also [107], where a more general algorithm can be found. For generalisations of the Panjer recursion, see also Chap. 4 of [47]. Section 1.10 follows Sect. 5 of [65]. More literature on premium principles may be found in [29, 71, 72, 88]. The premium principle of Example 1.15 was introduced in [135]. Convex risk measures, treated as premium principles, were first considered in [43]. Coherent risk measures were introduced in [7]. The generalisation to convex risk measures can be found in [61], see also the references therein. A discussion on risk measures in an actuarial context can be found in [45]. Further literature on the topic may be found, for example, in [47, 65, 70, 78, 88, 91, 110].

Chapter 2
Utility Theory

In this chapter we try to understand how agents take decisions. We assume that agents always act rationally. Then we determine the minimal premium an agent needs to take over a risk or the maximal premium an agent will accept to take insurance. Finally, we study the sharing of risk such that no agent can improve his/her position without deteriorating the position of another.

2.1 The Expected Utility Hypothesis

We consider an agent carrying a risk, for example a person who is looking for insurance, an insurance company looking for reinsurance, or a financial engineer with a certain portfolio. Alternatively, we could consider an agent who has to decide whether to take over a risk such as, for example, an insurance or a reinsurance company. We consider a simple one-period model. At the beginning of the period the agent has wealth w, at the end of the period he/she has wealth $w - X$, where X is the loss occurred from the risk. For simplicity we assume $X \ge 0$.

The agent is now offered an *insurance contract*. An insurance contract is specified by a *compensation function* $r : \mathbb{R}_+ \to \mathbb{R}_+$. The insurer covers the amount $r(X)$, the insured covers $X - r(X)$. For the compensation of the risk taken over by the insurer, the insured has to pay the *premium* π. Let us first consider the properties $r(x)$ should have. Because the insured is looking for security, the function $r(x)$ should be non-negative. Otherwise, the insured may take over a larger risk (for a possibly negative premium). On the other hand, there should be no over-compensation, i.e. $0 \le r(x) \le x$. Otherwise, the insured could make a gain from a loss. Let $s(x) = x - r(x)$ be the *self-insurance function*, the amount covered by the insured. Sometimes $s(x)$ is also called the *franchise*.

The insured now has the problem of finding the appropriate insurance cover. If the policy $(\pi, r(x))$ is used, the wealth will be $Y = w - \pi - s(X)$ after the period.

H. Schmidli, *Risk Theory*, Springer Actuarial,
DOI:10.1007/978-3-319-72005-0_2

A first idea would be to consider the expected wealth $\mathbb{E}[Y]$ and to choose the policy with the largest expected wealth. Because some administration costs will be included in the premium, the optimal decision will then always be not to insure the risk. The agent wants to reduce his risk, and therefore is interested in buying insurance. He therefore would be willing to pay a premium π such that $\mathbb{E}[Y] < \mathbb{E}[w-X]$, provided the risk is reduced. A second idea would be to compare the variances. Let Y_1 and Y_2 be two wealths corresponding to different insurance treaties. We prefer the first insurance form if $\text{Var}[Y_1] < \text{Var}[Y_2]$. This will only make sense if $\mathbb{E}[Y_1] = \mathbb{E}[Y_2]$. The disadvantage of the above approach is that a loss $\mathbb{E}[X] - x$ will be considered to be as bad as a loss $\mathbb{E}[X] + x$. Of course, the insured prefers a smaller to a larger loss.

The above discussion inspires the following concept. The agent gives a value to each possible wealth; i.e., he has a function $u : I \to \mathbb{R}$, called a *utility function*, giving a value to each possible wealth. Here I is the interval of possible wealth attainable by any insurance under consideration. I is assumed not to be a singleton. His criterion will now be *expected utility*; i.e., he will prefer insurance form one if $\mathbb{E}[u(Y_1)] > \mathbb{E}[u(Y_2)]$.

We now try to find properties a utility function should have. The agent will of course prefer larger wealth. Thus $u(x)$ must be *strictly increasing*. Moreover, the agent is risk averse, i.e. he will weight a loss higher for small wealth than for large wealth. Mathematically we can express this in the following way. For any $h > 0$ the function $y \mapsto u(y + h) - u(y)$ is *strictly decreasing*. Note that if $u(y)$ is a utility function fulfilling the above conditions, then also $\tilde{u}(y) = a + bu(y)$ for $a \in \mathbb{R}$ and $b > 0$ fulfils the conditions. Moreover, $\tilde{u}(y)$ and $u(y)$ will lead to the same decisions.

Let us consider some possible utility functions.

- Quadratic utility
$$u(y) = y - \frac{y^2}{2c}, \qquad y \le c,\ c > 0. \tag{2.1a}$$
- Exponential utility
$$u(y) = -\mathrm{e}^{-cy}, \qquad y \in \mathbb{R},\ c > 0. \tag{2.1b}$$
- Logarithmic utility
$$u(y) = \log(c + y), \qquad y > -c. \tag{2.1c}$$
- HARA utility
$$u(y) = (y + c)^\alpha \qquad y > -c,\ 0 < \alpha < 1, \tag{2.1d}$$
  or
$$u(y) = -(c - y)^\alpha \qquad y < c,\ \alpha > 1. \tag{2.1e}$$

The second condition for a utility function looks a bit complicated. We therefore want to find a nicer equivalent condition.

Theorem 2.1 *A strictly increasing function $u(y)$ is a utility function if and only if it is strictly concave.*

Proof Suppose $u(y)$ is a utility function. Let $x < z$. Then

$$u\Big(x + \frac{1}{2}(z - x)\Big) - u(x) > u(z) - u\Big(x + \frac{1}{2}(z - x)\Big),$$

implying

$$u\Big(\frac{1}{2}(x + z)\Big) > \frac{1}{2}\Big(u(x) + u(z)\Big). \tag{2.2}$$

We now fix x and $z > x$. Define the function $v : [0, 1] \to [0, 1]$,

$$\alpha \mapsto v(\alpha) = \frac{u((1-\alpha)x + \alpha z) - u(x)}{u(z) - u(x)}.$$

Note that $u(y)$ concave is equivalent to $v(\alpha) \geq \alpha$ for $\alpha \in (0, 1)$. By (2.2) $v(\frac{1}{2}) > \frac{1}{2}$. We first note that

$$\begin{aligned} v\Big(\frac{1}{2}(\alpha_1 + \alpha_2)\Big) &= \frac{u(\frac{1}{2}[\{(1-\alpha_1)x + \alpha_1 z\} + \{(1-\alpha_2)x + \alpha_2 z\}]) - u(x)}{u(z) - u(x)} \\ &> \frac{\frac{1}{2}u((1-\alpha_1)x + \alpha_1 z) + \frac{1}{2}u((1-\alpha_2)x + \alpha_2 z) - u(x)}{u(z) - u(x)} \\ &= \frac{1}{2}(v(\alpha_1) + v(\alpha_2)). \end{aligned}$$

Let $n \in \mathbb{N}$ and let $\alpha_{n,j} = j2^{-n}$. Assume $v(\alpha_{n,j}) > \alpha_{n,j}$ for $1 \leq j \leq 2^n - 1$. Then clearly $v(\alpha_{n+1,2j}) = v(\alpha_{n,j}) > \alpha_{n,j} = \alpha_{n+1,2j}$. By (2.2) and since $\frac{1}{2}(\alpha_{n,j} + \alpha_{n,j+1}) = \alpha_{n+1,2j+1}$,

$$v(\alpha_{n+1,2j+1}) > \frac{1}{2}(v(\alpha_{j,n}) + v(\alpha_{n,j+1})) > \frac{1}{2}(\alpha_{n,j} + \alpha_{n,j+1}) = \alpha_{n+1,2j+1}.$$

Now let α be arbitrary and let j_n such that $\alpha_{n,j_n} \leq \alpha < \alpha_{n,j_n+1}$, $\alpha_n = \alpha_{n,j_n}$. Then

$$v(\alpha) \geq v(\alpha_n) > \alpha_n.$$

Because n is arbitrary we have $v(\alpha) \geq \alpha$. This implies that $u(y)$ is concave. Now let $\alpha < \frac{1}{2}$. Then

$$
\begin{aligned}
u((1-\alpha)x+\alpha z) &= u((1-2\alpha)x+2\alpha\frac{1}{2}(x+z)) \\
&\geq (1-2\alpha)u(x)+2\alpha u(\frac{1}{2}(x+z)) \\
&> (1-2\alpha)u(x)+2\alpha\frac{1}{2}(u(x)+u(z)) = (1-\alpha)u(x)+\alpha u(z)\,.
\end{aligned}
$$

A similar argument applies if $\alpha > \frac{1}{2}$. The converse statement is left as an exercise. □

Next we would like to have a measure of how much an agent likes the risk. Assume that $u(y)$ is twice differentiable. Intuitively, the more concave a utility function is, the more it is sensible to risk. So a risk averse agent will have a utility function with $-u''(y)$ large, but $-u''(y)$ is not a good measure for this purpose. Let $\tilde{u}(y) = a + bu(y)$. Then $-\tilde{u}''(y) = -bu''(y)$ would give a different risk aversion, even though it leads to the same decisions. In order to get rid of b define the *risk aversion function*

$$
a(y) = -\frac{u''(y)}{u'(y)} = -\frac{\mathrm{d}}{\mathrm{d}y}\ln u'(y)\,.
$$

Intuitively, we would assume that a wealthy person can take more risk than a poor person. This would mean that $a(y)$ should be decreasing. For our examples we have that (2.1a) and (2.1e) give increasing risk aversion, (2.1c) and (2.1d) give decreasing risk aversion, and (2.1b) gives constant risk aversion.

2.2 The Zero Utility Premium

An agent has to decide whether to insure a risk or not. Let $(\pi, r(x))$ be an insurance treaty and assume that it is a non-trivial insurance, i.e. Var[r(X)] > 0. If the agent takes the insurance he will have the final wealth $w - \pi - s(X)$ and if he does not take the insurance his final wealth will be $w - X$. Considering his expected utility, the agent will take the insurance if and only if

$$
\mathbb{E}[u(w-\pi-s(X))] \geq \mathbb{E}[u(w-X)]\,.
$$

If equality holds, the agent is indifferent. So taking the insurance is possible. The left-hand side of the above inequality is a continuous and strictly decreasing function of π. The largest value that can be attained is for $\pi = 0$, $\mathbb{E}[u(w-s(X))] > \mathbb{E}[u(w-X)]$. The lowest possible value is difficult to obtain, because it depends on the choice of I. Let us assume that the lower bound is less or equal to $\mathbb{E}[u(w-X)]$. Then there exists a unique value $\pi_r(w)$ such that

$$
\mathbb{E}[u(w-\pi_r-s(X))] = \mathbb{E}[u(w-X)]\,.
$$

The premium π_r is the largest premium the agent is willing to pay for the insurance treaty $r(x)$. We call π_r the *zero utility premium.* In many situations the zero utility premium π_r will be larger then the "fair premium" $\mathbb{E}[r(X)]$.

Theorem 2.2 *If both $r(x)$ and $s(x)$ are increasing and* $\mathrm{Var}[r(X)] > 0$ *then* $\pi_r(w) > \mathbb{E}[r(X)]$.

Proof From the definition of a strictly concave function it follows that the function $v(y)$, $x \mapsto v(x) = u(w - x)$, is strictly concave. Define the functions $g_1(x) = s(x) + \mathbb{E}[r(X)]$ and $g_2(x) = x$. These two functions are increasing. The function $g_2(x) - g_1(x) = r(x) - \mathbb{E}[r(X)]$ is increasing. Thus there exists an x_0 such that $g_1(x) \geq g_2(x)$ for $x < x_0$ and $g_1(x) \leq g_2(x)$ for $x > x_0$. Note that $\mathbb{E}[g_2(x) - g_1(x)] = 0$. Thus the conditions of Corollary G.8 are fulfilled and

$$\begin{aligned}\mathbb{E}[u(w - \mathbb{E}[r(X)] - s(X))] &= \mathbb{E}[v(g_1(X))] > \mathbb{E}[v(g_2(X))] \\ &= \mathbb{E}[u(w - X)] = \mathbb{E}[u(w - \pi_r - s(X))] .\end{aligned}$$

Because $\pi \mapsto \mathbb{E}[u(w - \pi - s(X))]$ is a strictly decreasing function it follows that $\pi_r > \mathbb{E}[r(X)]$. □

Assume now that u is twice continuously differentiable. Then we find the following result.

Theorem 2.3 *Let $r(x) = x$. If the risk aversion function is decreasing (increasing), then the zero utility premium $\pi(w)$ is decreasing (increasing).*

Proof Consider the function $v(y) = u'(u^{-1}(y))$. Taking the derivative yields

$$v'(y) = \frac{u''(u^{-1}(y))}{u'(u^{-1}(y))} = -a(u^{-1}(y)) .$$

Because $y \mapsto u^{-1}(y)$ is increasing, this gives that $v'(y)$ is increasing (decreasing). Thus $v(y)$ is convex (concave).

We have

$$\pi(w) = w - u^{-1}(\mathbb{E}[u(w - X)]) \quad \text{and} \quad \pi'(w) = 1 - \frac{\mathbb{E}[u'(w - X)]}{u'(u^{-1}(\mathbb{E}[u(w - X)]))} .$$

Let now $a(y)$ be decreasing (increasing). By Jensen's inequality we have

$$u'(u^{-1}(\mathbb{E}[u(w - X)])) \leq (\geq)\, \mathbb{E}[u'(u^{-1}(u(w - X)))] = \mathbb{E}[u'(w - X)]$$

and therefore $\pi'(w) \leq (\geq)\, 0$. □

It is not always desirable that the decision depends on the initial wealth. We have the following result.

Lemma 2.4 *Let $u(x)$ be a utility function. Then the following are equivalent:*

(i) $u(x)$ is exponential, i.e. $u(x) = -A\mathrm{e}^{-cx} + B$ for some $A, c > 0$, $B \in \mathbb{R}$.
(ii) For all losses X and all compensation functions $r(x)$, the zero utility premium π_r does not depend on the initial wealth w.
(iii) For all losses X and for full reinsurance $r(x) = x$, the zero utility premium π_r does not depend on the initial wealth w.

Proof This follows directly from Lemma 1.12. □

2.3 Optimal Insurance

Let us now consider an agent carrying several risks $X_1, X_2, \ldots, X_n$. The total risk is then $X = X_1 + \cdots + X_n$. We consider here a compensation function $r(X_1, \ldots, X_n)$ and a self-insurance function $s(X_1, \ldots, X_n) = X - r(X_1, \ldots, X_n)$. We say an insurance treaty is *global* if $r(X_1, \ldots, X_n)$ depends on X only, or equivalently, $s(X_1, \ldots, X_n)$ depends on X only. Otherwise, we call an insurance treaty *local*. We next prove that, in some sense, the global treaties are optimal for the agent.

Theorem 2.5 (Pesonen–Ohlin) *If the premium depends on the pure net premium only, then to every local insurance treaty there exists a global treaty with the same premium but a higher expected utility.*

Proof Let $(\pi, r(x_1, \ldots, x_n))$ be a local treaty. Consider the compensation function $R(x) = \mathbb{E}[r(X_1, \ldots, X_n) \mid X = x]$. If the premium depends on the pure net premium only, then clearly $R(X)$ is sold for the same premium, because

$$\mathbb{E}[R(X)] = \mathbb{E}[\mathbb{E}[r(X_1, \ldots, X_n) \mid X]] = \mathbb{E}[r(X_1, \ldots, X_n)]\;.$$

By Jensen's inequality

$$\begin{aligned}\mathbb{E}[u(w - \pi - s(X_1, \ldots, X_n))] &= \mathbb{E}[\mathbb{E}[u(w - \pi - s(X_1, \ldots, X_n)) \mid X]] \\ &< \mathbb{E}[u(\mathbb{E}[w - \pi - s(X_1, \ldots, X_n) \mid X])] = \mathbb{E}[u(w - \pi - (X - R(X)))]\;.\end{aligned}$$

The strict inequality follows from the fact that $s(X_1, \ldots, X_n) \neq S(X)$, because otherwise the insurance treaty would be global, and from the strict concavity. □

Remark 2.6 Note that the result does not say that the global treaty constructed is a good treaty. It only says that for the investigation of good treaties we can restrict to global treaties. ■

Let us now find the optimal insurance treaty for the insured.

Theorem 2.7 (Arrow–Ohlin) *Assume the premium depends on the pure net premium only and let $(\pi, r(x))$ be an insurance treaty with $\mathbb{E}[r(X)] > 0$. Then there exists* *a unique deductible* $b > 0$ *such that the* excess of loss insurance $r_b(x) = (x - b)^+$ *has the same premium. Moreover, if* $\mathbb{P}[r(X) \neq r_b(X)] > 0$ *then the treaty* $(\pi, r_b(x))$ *yields a higher utility.*

Proof The function $(x - b)^+$ is continuous and decreasing in b. We have $\mathbb{E}[(X - 0)^+] \geq \mathbb{E}[r(X)] > 0 = \lim_{b\to\infty} \mathbb{E}[(X - b)^+]$. Thus there exists a unique $b > 0$ such that $\mathbb{E}[(X - b)^+] = \mathbb{E}[r(X)]$. Thus $r_b(x)$ yields the same premium. Assume now $\mathbb{P}[r(X) \neq r_b(X)] > 0$. Let $s_b(x) = x - r_b(x)$. For $y < b$ we have

$$\mathbb{P}[s_b(X) \leq y] = \mathbb{P}[X \leq y] \leq \mathbb{P}[s(X) \leq y] ,$$

and for $y > b$

$$\mathbb{P}[s_b(X) \leq y] = 1 \geq \mathbb{P}[s(X) \leq y] .$$

Note that $v(x) = u(w - \pi - x)$ is a strictly concave function. Thus Ohlin's lemma yields

$$\mathbb{E}[u(w - \pi - s_b(X))] = \mathbb{E}[v(s_b(X))] > \mathbb{E}[v(s(X))] = \mathbb{E}[u(w - \pi - s(X))] .$$

□

2.4 The Position of the Insurer

We now consider the situation from the point of view of an insurer. The discussion on how to take decisions also applies to the insurer. Assume therefore that the insurer has the utility function $\bar{u}(x)$. Denote the initial wealth of the insurer by $\overline{w}$. For an insurance treaty $(\pi, r(x))$ the expected utility of the insurer becomes

$$\mathbb{E}[\bar{u}(\overline{w} + \pi - r(X))] .$$

The *zero utility premium* $\overline{\pi}_r(\overline{w})$ of the insurer is the solution to

$$\mathbb{E}[\bar{u}(\overline{w} + \overline{\pi}_r(\overline{w}) - r(X))] = \bar{u}(\overline{w}) .$$

The insurer will take over the risk if $\pi \geq \overline{\pi}_r(\overline{w})$. By Jensen's inequality we have

$$\bar{u}(\overline{w}) = \mathbb{E}[\bar{u}(\overline{w} + \overline{\pi}_r(\overline{w}) - r(X))] < \bar{u}(\overline{w} + \overline{\pi}_r(\overline{w}) - \mathbb{E}[r(X)]) ,$$

which yields $\overline{\pi}_r(\overline{w}) - \mathbb{E}[r(X)] > 0$. An insurance contract can thus be signed by both parties if and only if $\overline{\pi}_r(\overline{w}) \leq \pi \leq \pi_r(w)$. It follows, as in Theorem 2.5 (replacing $s(x_1, \ldots, x_n)$ by $r(x_1, \ldots, x_n)$), that the insurer prefers global treaties to local treaties. From Theorem 2.7 (replacing $s(x)$ by $r(x)$) it follows that a first risk deductible is optimal for the insurer. Thus the interests of insurer and insured conflict.

Let us now consider insurance forms that take care of the interests of the insured. The insured would like that the larger the loss the larger should be the fraction covered by the insurance company. We call a compensation function $r(x)$ a *Vajda compensation function* if $r(x)/x$ is an increasing function of x. Because $r_b(x) = (x-b)^+$ is a Vajda compensation function it is clear how the insured would choose $r(x)$. Let us now consider what is optimal for the insurer.

Theorem 2.8 *Suppose the premium depends on the pure net premium only. For any Vajda insurance treaty $(\pi, r(x))$ with $\mathbb{E}[r(X)] > 0$ there exists a* proportional insurance treaty $r_k(x) = (1-k)x$ *with the same premium. Moreover, if $\mathbb{P}[r(X) \neq r_k(X)] > 0$ then $(\pi, r_k(x))$ yields a higher utility for the insurer.*

Proof The function $[0,1] \to \mathbb{R} : k \mapsto \mathbb{E}[(1-k)X]$ is decreasing and has its image in the interval $[0, \mathbb{E}[X]]$. Because $0 < \mathbb{E}[r(X)] \le \mathbb{E}[X]$ there is a unique k such that $\mathbb{E}[r_k(X)] = \mathbb{E}[r(X)]$. These two treaties then have the same premium. Suppose now that $\mathbb{P}[r(X) \neq r_k(X)] > 0$. Note that $r(x)$ and $r_k(x)$ are increasing. The difference

$$\frac{r(x) - r_k(x)}{x} = \Big(\frac{r(x)}{x} - (1-k)\Big)$$

is an increasing function in x. Thus there is an x_0 such that $r(x) \le r_k(x)$ for $x < x_0$ and $r(x) \ge r_k(x)$ for $x > x_0$. The function $v(x) = \bar{u}(\overline{w} + \pi - x)$ is strictly concave. By Corollary G.8 we have

$$\mathbb{E}[\bar{u}(\overline{w} + \pi - r_k(X))] = \mathbb{E}[v(r_k(X))] > \mathbb{E}[v(r(X))] = \mathbb{E}[\bar{u}(\overline{w} + \pi - r_k(X))]\,.$$

This proves the result. □

As from the point of view of the insured, it turns out that the zero utility premium is only independent of the initial wealth if the utility is exponential.

Proposition 2.9 *Let $\bar{u}(x)$ be a utility function. Then the following are equivalent:*

(i) $\bar{u}(x)$ is exponential, i.e. $\bar{u}(x) = -A\mathrm{e}^{-cx} + B$ for some $A, c > 0$, $B \in \mathbb{R}$.
(ii) For all losses X and all compensation functions $r(x)$, the zero utility premium $\overline{\pi}_r$ does not depend on the initial wealth $\overline{w}$.

Proof The proof is analogous to that of Lemma 2.4. □

2.5 Pareto-Optimal Risk Exchanges

Consider now a market with n agents, i.e. insured, insurance companies, reinsurers. In the market there are the risks $\boldsymbol{X} = (X_1, \dots, X_n)^\top$. Some of the risks may be zero if the corresponding agent does not initially carry a risk. Here we also allow for financial risks (investment). If X_i is positive, we consider it as a loss and if it is negative, we consider it as a gain. Making contracts, the agents redistribute the risk $\boldsymbol{X}$. Formally, a *risk exchange* is a function

$$\boldsymbol{f} = (f_1, \dots, f_n)^\top : \mathbb{R}^n \to \mathbb{R}^n$$

such that

$$\sum_{i=1}^n f_i(\boldsymbol{X}) = \sum_{i=1}^n X_i \ . \tag{2.3}$$

The latter condition assures that the whole risk is covered. A risk exchange means that agent i covers $f_i(\boldsymbol{X})$. This is a generalisation of the insurance market we had considered before.

Now each of the agents has an initial wealth w_i and a utility function $u_i(x)$. Under the risk exchange $\boldsymbol{f}$ the expected terminal utility of agent i becomes

$$V_i(\boldsymbol{f}) = \mathbb{E}[u_i(w_i - f_i(\boldsymbol{X}))] \ .$$

As we saw before, the interests of the agents will conflict. Thus it will in general not be possible to find $\boldsymbol{f}$ such that each agent gets an optimal utility. However, an agent may agree to a risk exchange $\boldsymbol{f}$ if $V_i(\boldsymbol{f}) \geq \mathbb{E}[u_i(w_i - X_i)]$. Hence, which risk exchange will be chosen depends upon negotiations. However, the agents will agree not to discuss certain risk exchanges.

Definition 2.10 A risk exchange $\boldsymbol{f}$ is called *Pareto-optimal* if for any risk exchange $\tilde{\boldsymbol{f}}$ with

$$V_i(\boldsymbol{f}) \leq V_i(\tilde{\boldsymbol{f}})\ \forall i \quad \text{it follows that} \quad V_i(\boldsymbol{f}) = V_i(\tilde{\boldsymbol{f}})\ \forall i \ .$$

For a risk exchange $\boldsymbol{f}$ that is not Pareto-optimal we would have a risk exchange $\tilde{\boldsymbol{f}}$ that is at least as good for all agents, but better for at least one of them. Of course, it is not excluded that $V_i(\boldsymbol{f}) < V_i(\tilde{\boldsymbol{f}})$ for some i. If the latter is the case then there must be a j such that $V_j(\boldsymbol{f}) > V_j(\tilde{\boldsymbol{f}})$.

For the negotiations, the agents can in principle restrict to Pareto-optimal risk exchanges. Of course, in reality, the agents do not know the utility functions of the others. So the concept of Pareto optimality is more a theoretical way to consider the market.

It seems quite difficult to decide whether a certain risk exchange is Pareto-optimal or not. It turns out that a simple criteria does the job.

Theorem 2.11 (Borch/du Mouchel) *Assume that the utility functions $u_i(x)$ are differentiable. A risk exchange $\boldsymbol{f}$ is Pareto-optimal if and only if there exist numbers θ_i such that (almost surely)*

$$u_i'(w_i - f_i(\boldsymbol{X})) = \theta_i u_1'(w_1 - f_1(\boldsymbol{X})) \ . \tag{2.4}$$

Remark 2.12 Note that necessarily $\theta_1 = 1$ and $\theta_i > 0$ because $u_i'(x) > 0$. Note also that we could make the comparisons with agent j instead of agent 1. Then we just need to divide the θ_i by θ_j, for example θ_1 would be replaced by θ_j^{-1}. ■

Proof Assume that θ_i exist such that (2.4) is satisfied. Let $\tilde{\boldsymbol{f}}$ be a risk exchange such that $V_i(\tilde{\boldsymbol{f}}) \geq V_i(\boldsymbol{f})$ for all i. By concavity

$$u_i(w_i - \tilde{f}_i(\boldsymbol{X})) \leq u_i(w_i - f_i(\boldsymbol{X})) + u_i'(w_i - f_i(\boldsymbol{X}))(f_i(\boldsymbol{X}) - \tilde{f}_i(\boldsymbol{X}))\,.$$

This yields

$$\frac{u_i(w_i - \tilde{f}_i(\boldsymbol{X})) - u_i(w_i - f_i(\boldsymbol{X}))}{\theta_i} \leq u_1'(w_1 - f_1(\boldsymbol{X}))(f_i(\boldsymbol{X}) - \tilde{f}_i(\boldsymbol{X}))\,.$$

Using $\sum_{i=1}^n f_i(\boldsymbol{X}) = \sum_{i=1}^n X_i = \sum_{i=1}^n \tilde{f}_i(\boldsymbol{X})$ we find by summing over all i,

$$\sum_{i=1}^n \frac{u_i(w_i - \tilde{f}_i(\boldsymbol{X})) - u_i(w_i - f_i(\boldsymbol{X}))}{\theta_i} \leq 0\,.$$

Taking expectations yields

$$\sum_{i=1}^n \frac{V_i(\tilde{\boldsymbol{f}}) - V_i(\boldsymbol{f})}{\theta_i} \leq 0\,.$$

From the assumption $V_i(\tilde{\boldsymbol{f}}) \geq V_i(\boldsymbol{f})$ it follows that $V_i(\tilde{\boldsymbol{f}}) = V_i(\boldsymbol{f})$. Thus $\boldsymbol{f}$ is Pareto-optimal.

Assume now that (2.4) does not hold. By renumbering, we can assume that there is no constant θ_2 such that (2.4) holds for $i = 2$. Let

$$\theta = \frac{\mathbb{E}[u_1'(w_1 - f_1(\boldsymbol{X}))u_2'(w_2 - f_2(\boldsymbol{X}))]}{\mathbb{E}[(u_1'(w_1 - f_1(\boldsymbol{X})))^2]}$$

and $W = u_2'(w_2 - f_2(\boldsymbol{X})) - \theta u_1'(w_1 - f_1(\boldsymbol{X}))$. We then have $\mathbb{E}[W u_1'(w_1 - f_1(\boldsymbol{X}))] = 0$ and by the assumption $\mathbb{E}[W^2] > 0$. Let now $\varepsilon > 0$, $\delta = \frac{1}{2}\mathbb{E}[W^2]/\mathbb{E}[u_2'(w_2 - f_2(\boldsymbol{X}))] > 0$ and $\tilde{f}_1(\boldsymbol{X}) = f_1(\boldsymbol{X}) - (\delta - W)\varepsilon$, $\tilde{f}_2(\boldsymbol{X}) = f_2(\boldsymbol{X}) + (\delta - W)\varepsilon$, $\tilde{f}_i(\boldsymbol{X}) = f_i(\boldsymbol{X})$ for all $i > 2$. Then $\tilde{\boldsymbol{f}}$ is a risk exchange. Consider now

$$\begin{aligned}
&\frac{V_1(\tilde{\boldsymbol{f}}) - V_1(\boldsymbol{f})}{\varepsilon} - \mathbb{E}[u_1'(w_1 - f_1(\boldsymbol{X}))]\delta \\
&\quad = \mathbb{E}\Big[(\delta - W)\Big(\frac{u_1(w_1 - \tilde{f}_1(\boldsymbol{X})) - u_1(w_1 - f_1(\boldsymbol{X}))}{(\delta - W)\varepsilon} - u_1'(w_1 - f_1(\boldsymbol{X}))\Big)\Big],
\end{aligned}$$

where we used $\mathbb{E}[Wu_1'(w_1 - f_1(\boldsymbol{X}))] = 0$. The right-hand side tends to zero as ε tends to zero. Thus $V_1(\tilde{\boldsymbol{f}}) > V_1(\boldsymbol{f})$ for ε small enough. We also have

$$\frac{V_2(\tilde{\boldsymbol{f}}) - V_2(\boldsymbol{f})}{\varepsilon} - \mathbb{E}[u_2'(w_2 - f_2(\boldsymbol{X}))(W - \delta)]$$
$$= \mathbb{E}\Big[(W - \delta)\Big(\frac{u_2(w_2 - \tilde{f}_2(\boldsymbol{X})) - u_2(w_2 - f_2(\boldsymbol{X}))}{(W - \delta)\varepsilon} - u_2'(w_2 - f_2(\boldsymbol{X}))\Big)\Big] .$$

Also here the right-hand side tends to zero. We have

$$\begin{aligned}
&\mathbb{E}[u_2'(w_2 - f_2(\boldsymbol{X}))(W - \delta)] \\
&\quad = \mathbb{E}[(W + \theta u_1'(w_1 - f_1(\boldsymbol{X})))W] - \delta\mathbb{E}[u_2'(w_2 - f_2(\boldsymbol{X}))] \\
&\quad = \mathbb{E}[W^2] - \delta\mathbb{E}[u_2'(w_2 - f_2(\boldsymbol{X}))] = \frac{1}{2}\mathbb{E}[W^2] > 0 .
\end{aligned}$$

Thus also $V_2(\tilde{\boldsymbol{f}}) > V_2(\boldsymbol{f})$ for ε small enough. This shows that $\boldsymbol{f}$ is not Pareto-optimal. □

In order to find the Pareto-optimal risk exchanges one has to solve Eq. (2.4) subject to the constraint (2.3).

If a risk exchange is chosen, the quantity $f_i(0, \dots, 0)$ is the amount agent i has to pay (obtains if negative) if no losses occur. Thus $f_i(0, \dots, 0)$ can be interpreted as a premium.

Note that for Pareto-optimality only the support of the distribution of $\boldsymbol{X}$, not the distribution itself, has an influence. Hence the Pareto-optimal solution can be found without investigating the risk distribution, or the dependencies. It also shows that any Pareto optimal solution will be a function of $\sum_{i=1}^n X_i$ only.

We now solve the problem explicitly. Because $u_i'(x)$ is strictly decreasing, it is invertible and its inverse $(u_i')^{-1}(y)$ is strictly decreasing. Moreover, if $u_i'(x)$ is continuous (as it is under the conditions of Theorem 2.11), then $(u_i')^{-1}(y)$ is continuous, too. Choose $\theta_i > 0$, $\theta_1 = 1$. Then (2.4) yields

$$w_i - f_i(\boldsymbol{X}) = (u_i')^{-1}(\theta_i u_1'(w_1 - f_1(\boldsymbol{X}))) .$$

Summing over i gives by use of (2.3)

$$\sum_{i=1}^n w_i - \sum_{i=1}^n X_i = \sum_{i=1}^n (u_i')^{-1}(\theta_i u_1'(w_1 - f_1(\boldsymbol{X}))) .$$

The function

$$g(y) = \sum_{i=1}^n (u_i')^{-1}(\theta_i u_1'(y))$$

is then strictly increasing and therefore invertible. If all the $u_i'(x)$ are continuous, then also $g(y)$ is continuous and therefore $g^{-1}(x)$ is continuous, too. This yields

$$f_1(\boldsymbol{X}) = w_1 - g^{-1}\Bigl(\sum_{i=1}^{n} w_i - \sum_{i=1}^{n} X_i\Bigr).$$

For i arbitrary we then obtain

$$f_i(\boldsymbol{X}) = w_i - (u_i')^{-1}\bigl(\theta_i u_1'\bigl(g^{-1}\bigl(\sum_{i=1}^{n} w_i - \sum_{i=1}^{n} X_i\bigr)\bigr)\bigr).$$

Thus we have found the following result.

Theorem 2.13 *A Pareto-optimal risk exchange is a pool, that is, each individual agent contributes a share that depends on the total loss of the group only. Moreover, each agent must cover a genuine part of any increase of the total loss of the group. If all utility functions are differentiable, then the individual shares are continuous functions of the total loss.* □

Remark 2.14 Consider the situation insured-insurer, i.e. $n = 2$ and a risk of the form $(X, 0)^\top$, where $X \geq 0$. For certain choices of θ_2 it is possible that $f_1(0, 0) < 0$, i.e. the insurer has to pay a premium to take over the risk. This shows that not all possible choices of θ_i will be realistic. In fact, the risk exchange has to be chosen such that the expected utility of each agent is increased, i.e. $\mathbb{E}[u_i(w_i - f_i(\boldsymbol{X}))] \geq \mathbb{E}[u_i(w_i - X_i)]$ in order that agent i will be willing to participate. ■

Bibliographical Remarks

Utility theory goes back to Nicholas Bernoulli's St. Petersburg paradox. It was formalised in [134]. More results are found in [6]. The results presented here can be found in [123]. Optimal insurance markets were studied, for example, in [23, 24, 30, 50, 64]. Further literature on the topic may be found, for example, in [47, 78].

Chapter 3
Credibility Theory

Bayesian methods are introduced for the calculation of the net premium of a collective insurance contract. Because the joint distributions are hard to obtain, one uses a linear prediction of the aggregate claims. We will further consider Bonus-Malus systems as a simple approximation to the credibility premium.

3.1 Introduction

We start with an example. A company has insured twenty drivers in a small village during the past ten years. Table 3.1 shows, indicated by a 1, which driver had at least one accident in the corresponding year. The last row shows the number of years with at least one claim. Some of the drivers had no accidents. They will think that it is unfair that they have to pay the same premium as the drivers with four or more years with accidents. Another insurer could offer the good drivers a cheaper contract, and the first insurer would lose them. Thus the first insurer should offer the good risks a cheaper premium, and therefore has to increase the premia for the poor risks.

Could it not be possible that the probability of at least one accident in a year is the same for all the drivers? Had some of the drivers just been unlucky? We can test this by using a χ^2-test. Let $\hat{\mu}_i$ be the mean number of years with an accident for driver i and

$$\bar{\mu} = \frac{1}{20}\sum_{i=1}^{20} \hat{\mu}_i$$

be the mean number of years with an accident for a typical driver. Then

$$Z = \frac{10\sum_{i=1}^{20}(\hat{\mu}_i - \bar{\mu})^2}{\bar{\mu}(1-\bar{\mu})} = 49.1631\,.$$

H. Schmidli, *Risk Theory*, Springer Actuarial,
DOI:10.1007/978-3-319-72005-0_3

Table 3.1 Years with accidents

	1	2	3	4	5	6	7	8	9	10	11	12	13	14	15	16	17	18	19	20
1			1							1							1	1		
2							1		1	1							1			
3			1				1		1				1							
4																	1			
5									1	1										
6						1			1		1									
7									1					1			1			
8											1									
9						1				1		1								
10									1		1						1		1	
Σ	0	0	2	0	0	2	2	0	6	4	3	1	1	1	0	0	5	1	1	0

Under the hypothesis that $\mathbb{E}[\mu_i]$ is the same for all drivers, the statistic Z is approximately χ^2_{19}-distributed. However,

$$\mathbb{P}[\chi^2_{19} > 49] \approx 0.0002\,.$$

Thus it is very unlikely that each driver has the same probability of having at least one accident in a year.

For the insurance company it is preferable to attract good risks and to get rid of the poor risks. Thus they would like to charge premia according to their experience with a particular customer. This procedure is called *experience rating* or *credibility*. Let us denote by Y_{ij} the losses of the i-th risk in year j, where $i = 1, 2, \ldots, n$ and $j = 1, 2, \ldots, m$. Denote the mean losses of the i-th risk by

$$\bar{Y}_i := \frac{1}{m}\sum_{j=1}^{m} Y_{ij}\,.$$

We make the following assumptions:

(i) There exists a parameter Θ_i belonging to the risk i. The parameters $(\Theta_i : 1 \le i \le n)$ are iid.
(ii) The vectors $((Y_{i1}, Y_{i2}, \ldots, Y_{im}, \Theta_i) : 1 \le i \le n)$ are iid.
(iii) For fixed i, given Θ_i the random variables $Y_{i1}, Y_{i2}, \ldots, Y_{im}$ are conditionally iid.

For instance, Θ_i can be the parameters of a family of distributions, or more generally, the distribution function of a claim of the i-th risk.

Denote by $m(\vartheta) = \mathbb{E}[Y_{ij} \mid \Theta_i = \vartheta]$ the conditional expected value of an aggregate loss given $\Theta_i = \vartheta$, by $\mu = \mathbb{E}[m(\Theta_i)]$ the overall mean of the aggregate losses and by $s^2(\vartheta) = \operatorname{Var}[Y_{ij} \mid \Theta_i = \vartheta]$ the conditional variance of Y_{ij} given

$\Theta_i = \vartheta$. Now instead of using the expected value μ of a claim of a typical risk for the premium calculations one should use $m(\Theta_i)$ instead. Unfortunately we do not know Θ_i. Although $\bar{Y}_i$ is an estimator for $m(\Theta_i)$, it is not a good idea to use it for the premium calculation. Assume that somebody who is insured for the first year has a big claim. His premium for the next year would become larger than his claim and he would no longer be able to take insurance. However, the idea behind insurance is that the risk is shared over all the insurance takers, ensuring that nobody gets into trouble just because he is unlucky. Thus we have to find other methods to estimate $m(\Theta_i)$.

3.2 Bayesian Credibility

Suppose we know the distribution of Θ_i and the conditional distribution of Y_{ij} given Θ_i. We now want to find the best estimate M_0 of $m(\Theta_i)$ in the sense that

$$\mathbb{E}[(M - m(\Theta_i))^2] \geq \mathbb{E}[(M_0 - m(\Theta_i))^2]$$

for all measurable functions $M = M(Y_{i1}, Y_{i2}, \dots, Y_{im})$. Let $M_0 = \mathbb{E}[m(\Theta_i) \mid Y_{i1}, Y_{i2}, \dots, Y_{im}]$. Then

$$\begin{aligned}\mathbb{E}[(M - m(\Theta_i))^2] &= \mathbb{E}[\{(M - M_0) + (M_0 - m(\Theta_i))\}^2] \\ &= \mathbb{E}[(M - M_0)^2] + \mathbb{E}[(M_0 - m(\Theta_i))^2] + 2\mathbb{E}[(M - M_0)(M_0 - m(\Theta_i))] \,.\end{aligned}$$

We then have

$$\begin{aligned}&\mathbb{E}[(M - M_0)(M_0 - m(\Theta_i))] \\ &\quad = \mathbb{E}[\mathbb{E}[(M - M_0)(M_0 - m(\Theta_i)) \mid Y_{i1}, Y_{i2}, \dots, Y_{im}]] \\ &\quad = \mathbb{E}[(M - M_0)(M_0 - \mathbb{E}[m(\Theta_i) \mid Y_{i1}, Y_{i2}, \dots, Y_{im}])] \\ &\quad = 0 \,.\end{aligned}$$

Thus

$$\mathbb{E}[(M - m(\Theta_i))^2] = \mathbb{E}[(M - M_0)^2] + \mathbb{E}[(M_0 - m(\Theta_i))^2] \,.$$

The second term is independent of the choice of M. This means that the best estimate is M_0. We call M_0 the Bayesian credibility estimator. In the sequel we will consider the two most often used Bayesian credibility models.

3.2.1 The Poisson-Gamma Model

We assume that Y_{ij} has a compound Poisson distribution with parameter λ_i and an individual claim size distribution which is the same for all risks. It is therefore sufficient for the company to estimate λ_i. In the sequel we assume that all claim sizes are equal to 1, i.e. $Y_{ij} = N_{ij}$. Motivated by the compound negative binomial model we assume that $\lambda_i \sim \Gamma(\gamma, \alpha)$. Let

$$\bar{N}_i = \frac{1}{m}\sum_{j=1}^{m} N_{ij}$$

denote the mean number of claims of the i-th risk. We know the overall mean $\mathbb{E}[\lambda_i] = \gamma\alpha^{-1}$. For the best estimator for λ_i we get

$$\mathbb{E}[\lambda_i \mid N_{i1}, N_{i2}, \ldots, N_{im}] = \frac{\displaystyle\int_0^\infty \ell \prod_{j=1}^{m}\Big(\frac{\ell^{N_{ij}}}{N_{ij}!}\mathrm{e}^{-\ell}\Big)\frac{\alpha^\gamma}{\Gamma(\gamma)}\ell^{\gamma-1}\mathrm{e}^{-\alpha\ell}\,\mathrm{d}\ell}{\displaystyle\int_0^\infty \prod_{j=1}^{m}\Big(\frac{\ell^{N_{ij}}}{N_{ij}!}\mathrm{e}^{-\ell}\Big)\frac{\alpha^\gamma}{\Gamma(\gamma)}\ell^{\gamma-1}\mathrm{e}^{-\alpha\ell}\,\mathrm{d}\ell}$$

$$= \frac{\displaystyle\int_0^\infty \ell^{m\bar{N}_i+\gamma}\mathrm{e}^{-(m+\alpha)\ell}\,\mathrm{d}\ell}{\displaystyle\int_0^\infty \ell^{m\bar{N}_i+\gamma-1}\mathrm{e}^{-(m+\alpha)\ell}\,\mathrm{d}\ell} = \frac{\Gamma(m\bar{N}_i+\gamma+1)(m+\alpha)^{m\bar{N}_i+\gamma}}{(m+\alpha)^{m\bar{N}_i+\gamma+1}\Gamma(m\bar{N}_i+\gamma)}$$

$$= \frac{m\bar{N}_i+\gamma}{m+\alpha} = \frac{m}{m+\alpha}\bar{N}_i + \frac{\alpha}{m+\alpha}\Big(\frac{\gamma}{\alpha}\Big)$$

$$= \frac{m}{m+\alpha}\bar{N}_i + \Big(1-\frac{m}{m+\alpha}\Big)\mathbb{E}[\lambda_i].$$

The best estimator is therefore of the form

$$Z\bar{N}_i + (1-Z)\mathbb{E}[\lambda_i]$$

with the *credibility factor*

$$Z = \frac{1}{1+\dfrac{\alpha}{m}}.$$

The credibility factor Z is increasing with the number of years m and converges to 1 as $m \to \infty$. Thus the more observations we have, the higher the weight given to the empirical mean $\bar{N}_i$.

The only problem that remains is to estimate the parameters γ and α. For an insurance company this is usually not a problem because they have big data sets.

3.2.2 The Normal-Normal Model

For the next model we do not assume that the individual claims have the same distribution for all risks. First we assume that each risk consists of a lot of sub-risks such that the random variables Y_{ij} are approximately normally distributed. Moreover, the conditional variance is assumed to be the same for all risks, i.e. $Y_{ij} \sim \mathrm{N}(\Theta_i, \sigma^2)$. Next we assume that the parameter $\Theta_i \sim \mathrm{N}(\mu, \eta^2)$ is also normally distributed. Note that μ and η^2 have to be chosen in such a way that $\mathbb{P}[Y_{ij} \le 0]$ is small.

Let us first consider the joint density of $Y_{i1}, Y_{i2}, \dots Y_{im}, \Theta_i$

$$(2\pi)^{-(m+1)/2}\eta^{-1}\sigma^{-m} \exp\Bigl\{-\Bigl(\frac{(\vartheta-\mu)^2}{2\eta^2} + \sum_{j=1}^m \frac{(y_{ij}-\vartheta)^2}{2\sigma^2}\Bigr)\Bigr\}.$$

Because we will mainly be interested in the posterior distribution of Θ_i we should write the exponent in the form $(\vartheta - \cdot)^2$. We will use the abbreviation $m\bar{y}_i = \sum_{j=1}^m y_{ij}$. Then

$$\frac{(\vartheta-\mu)^2}{2\eta^2} + \sum_{j=1}^m \frac{(y_{ij}-\vartheta)^2}{2\sigma^2} = \vartheta^2\Bigl(\frac{1}{2\eta^2} + \frac{m}{2\sigma^2}\Bigr) - 2\vartheta\Bigl(\frac{\mu}{2\eta^2} + \frac{m\bar{y}_i}{2\sigma^2}\Bigr) + \frac{\mu^2}{2\eta^2} + \sum_{j=1}^m \frac{y_{ij}^2}{2\sigma^2}.$$

The joint density can therefore be written as

$$C(y_{i1}, y_{i2}\cdots, y_{im}) \exp\left\{-\frac{\Bigl[\vartheta - \Bigl(\frac{\mu}{\eta^2} + \frac{m\bar{y}_i}{\sigma^2}\Bigr)\Bigl(\frac{1}{\eta^2} + \frac{m}{\sigma^2}\Bigr)^{-1}\Bigr]^2}{2\Bigl(\frac{1}{\eta^2} + \frac{m}{\sigma^2}\Bigr)^{-1}}\right\}$$

where C is a function of the data. Hence the posterior distribution of Θ_i is normally distributed with expected value

$$\Bigl(\frac{\mu}{\eta^2} + \frac{m\bar{Y}_i}{\sigma^2}\Bigr)\Bigl(\frac{1}{\eta^2} + \frac{m}{\sigma^2}\Bigr)^{-1} = \frac{\mu\sigma^2 + m\eta^2\bar{Y}_i}{\sigma^2 + m\eta^2} = \frac{m\eta^2}{\sigma^2 + m\eta^2}\bar{Y}_i + \Bigl(1 - \frac{m\eta^2}{\sigma^2 + m\eta^2}\Bigr)\mu.$$

Again the credibility premium is a linear combination $Z\bar{Y}_i + (1-Z)\mathbb{E}[Y_{ij}]$ of the mean losses of the i-th risk and the overall mean μ. The credibility factor has a similar form as in the Poisson-gamma model

$$Z = \frac{1}{1 + \dfrac{\sigma^2}{\eta^2 m}}.$$

3.2.3 *Is the Credibility Premium Formula Always Linear?*

From the previous considerations one might conjecture that the credibility premium is always of the form $Z\bar{Y}_i + (1-Z)\mathbb{E}[Y_{ij}]$. We try to find a counterexample. Assume that Θ_i takes the values 1 and 2 with equal probabilities $\frac{1}{2}$. Given Θ_i the random variables $Y_{ij} \sim \text{Pois}(\Theta_i)$ are Poisson distributed. Then

$$\begin{aligned}
&\mathbb{E}[\Theta_i \mid Y_{i1}, Y_{i2}, \ldots, Y_{im}] \\
&\quad = \mathbb{P}[\Theta_i = 1 \mid Y_{i1}, Y_{i2}, \ldots, Y_{im}] + 2\mathbb{P}[\Theta_i = 2 \mid Y_{i1}, Y_{i2}, \ldots, Y_{im}] \\
&\quad = 2 - \mathbb{P}[\Theta_i = 1 \mid Y_{i1}, Y_{i2}, \ldots, Y_{im}] = 2 - \frac{\frac{1}{2}\prod_{j=1}^{m} \frac{1}{Y_{ij}!}\mathrm{e}^{-1}}{\frac{1}{2}\Bigl(\prod_{j=1}^{m} \frac{1}{Y_{ij}!}\mathrm{e}^{-1} + \prod_{j=1}^{m} \frac{2^{Y_{ij}}}{Y_{ij}!}\mathrm{e}^{-2}\Bigr)} \\
&\quad = 2 - \frac{1}{1 + \mathrm{e}^{-m}2^{m\bar{Y}_i}}\,.
\end{aligned}$$

Thus, in this case, we cannot get a linear formula.

3.3 Empirical Bayes Credibility

In order to calculate the Bayes credibility estimator we need to know the joint distribution of $m(\Theta_i)$ and $(Y_{i1}, Y_{i,2}, \ldots, Y_{im})$. In practice we do not have distributions but only data. It therefore does not seem feasible that we can estimate joint distributions if only one of the variables is observable. Thus we want to restrict now to linear estimators, i.e. estimators of the form

$$M = a_{i0} + a_{i1}Y_{i1} + a_{i2}Y_{i2} + \cdots + a_{im}Y_{im}\,.$$

The best estimator is called the *linear Bayes estimator*. It turns out that it is enough to know the first two moments of $(m(\Theta_i), Y_{i1})$. One then has to estimate these quantities. Furthermore, we do not know the expected values or the variances. In practice we will have to estimate these quantities. We will therefore proceed in two steps: First we estimate the linear Bayes premium, and then we estimate the first two moments. The corresponding estimator is called the *empirical Bayes estimator*.

3.3.1 The Bühlmann Model

In addition to the notation introduced earlier, let $\sigma^2 = \mathbb{E}[s^2(\Theta_i)]$ and $v^2 = \text{Var}[m(\Theta_i)]$. Note that σ^2 is not the unconditional variance of Y_{ij}. In fact

$$\mathbb{E}[Y_{ij}^2] = \mathbb{E}[\mathbb{E}[Y_{ij}^2 \mid \Theta_i]] = \mathbb{E}[s^2(\Theta_i) + (m(\Theta_i))^2] = \sigma^2 + \mathbb{E}[(m(\Theta_i))^2] .$$

Thus the variance of Y_{ij} is

$$\text{Var}[Y_{ij}] = \sigma^2 + \mathbb{E}[(m(\Theta_i))^2] - \mu^2 = \sigma^2 + v^2 . \tag{3.1}$$

We consider credibility premia of the form

$$a_{i0} + \sum_{j=1}^{m} a_{ij} Y_{ij} .$$

Which parameters a_{ij} minimise the expected quadratic error

$$\mathbb{E}\Bigl[\Bigl(a_{i0} + \sum_{j=1}^{m} a_{ij} Y_{ij} - m(\Theta_i)\Bigr)^2\Bigr] ?$$

We first differentiate with respect to a_{i0}.

$$\begin{aligned}\frac{1}{2}\frac{\mathrm{d}}{\mathrm{d}a_{i0}}\mathbb{E}\Bigl[\Bigl(a_{i0} + \sum_{j=1}^{m} a_{ij} Y_{ij} - m(\Theta_i)\Bigr)^2\Bigr] &= \mathbb{E}\Bigl[a_{i0} + \sum_{j=1}^{m} a_{ij} Y_{ij} - m(\Theta_i)\Bigr] \\ &= a_{i0} + \Bigl(\sum_{j=1}^{m} a_{ij} - 1\Bigr)\mu = 0 .\end{aligned}$$

It follows that

$$a_{i0} = \Bigl(1 - \sum_{j=1}^{m} a_{ij}\Bigr)\mu$$

and thus our estimator has the property that

$$\mathbb{E}\Bigl[a_{i0} + \sum_{j=1}^{m} a_{ij} Y_{ij}\Bigr] = \mu .$$

Let $1 \le k \le m$ and differentiate with respect to a_{ik}. Then

$$0 = \frac{1}{2}\frac{\mathrm{d}}{\mathrm{d}a_{ik}}\mathbb{E}\Big[\Big(a_{i0} + \sum_{j=1}^{m} a_{ij}Y_{ij} - m(\Theta_i)\Big)^2\Big]$$
$$= \mathbb{E}\Big[Y_{ik}\Big(a_{i0} + \sum_{j=1}^{m} a_{ij}Y_{ij} - m(\Theta_i)\Big)\Big]$$
$$= a_{i0}\mu + \sum_{\substack{j=1\\ j\neq k}}^{m} a_{ij}\mathbb{E}[Y_{ik}Y_{ik}] + a_{ik}\mathbb{E}[Y_{ik}^2] - \mathbb{E}[Y_{ik}m(\Theta_i)]\,.$$

We already know from (3.1) that $\mathbb{E}[Y_{ik}^2] = \sigma^2 + v^2 + \mu^2$. For $j \neq k$ we get

$$\mathbb{E}[Y_{ik}Y_{ij}] = \mathbb{E}[\mathbb{E}[Y_{ik}Y_{ij} \mid \Theta_i]] = \mathbb{E}[(m(\Theta_i))^2] = v^2 + \mu^2\,.$$

Finally,

$$\mathbb{E}[Y_{ik}m(\Theta_i)] = \mathbb{E}[\mathbb{E}[Y_{ik}m(\Theta_i) \mid \Theta_i]] = \mathbb{E}[(m(\Theta_i))^2] = v^2 + \mu^2\,.$$

The equations to solve are

$$\Big(1 - \sum_{j=1}^{m} a_{ij}\Big)\mu^2 + \sum_{\substack{j=1\\ j\neq k}}^{m} a_{ij}(v^2 + \mu^2) + a_{ik}(\sigma^2 + v^2 + \mu^2) - (v^2 + \mu^2)$$
$$= a_{ik}\sigma^2 - \Big(1 - \sum_{j=1}^{m} a_{ij}\Big)v^2 = 0\,.$$

The right-hand side of

$$\sigma^2 a_{ik} = v^2\Big(1 - \sum_{j=1}^{m} a_{ij}\Big)$$

is independent of k. Thus

$$a_{i1} = a_{i2} = \cdots = a_{im} = \frac{v^2}{\sigma^2 + mv^2}\,. \tag{3.2}$$

The credibility premium is

$$\frac{mv^2}{\sigma^2 + mv^2}\bar{Y}_i + \Big(1 - \frac{mv^2}{\sigma^2 + mv^2}\Big)\mu = Z\bar{Y}_i + (1 - Z)\mu\,, \tag{3.3}$$

where

$$Z = \frac{1}{1 + \dfrac{\sigma^2}{v^2 m}}\,.$$

The formula for the credibility premium only depends on μ, σ^2, and v^2. These were the only quantities we had assumed in our model. Hence the approach is quite general. We do not need any further assumptions on the distribution of Θ_i.

In order to apply the result we need to estimate the parameters μ, σ^2, and v^2. In order to make the right decision we look for unbiased estimators.

(i) $\boldsymbol{\mu}$: The natural estimator of μ is

$$\hat{\mu} = \frac{1}{nm}\sum_{i=1}^{n}\sum_{j=1}^{m} Y_{ij} = \frac{1}{n}\sum_{i=1}^{n}\bar{Y}_i \,. \tag{3.4}$$

It is easy to see that $\hat{\mu}$ is unbiased. It can be shown that $\hat{\mu}$ is the best linear unbiased estimator.

(ii) $\boldsymbol{\sigma}^2$: For the estimation of $s^2(\Theta_i)$ we would use the unbiased estimator

$$\frac{1}{m-1}\sum_{j=1}^{m}(Y_{ij}-\bar{Y}_i)^2 \,.$$

Thus

$$\hat{\sigma}^2 = \frac{1}{n(m-1)}\sum_{i=1}^{n}\sum_{j=1}^{m}(Y_{ij}-\bar{Y}_i)^2 \tag{3.5}$$

is an unbiased estimator for σ^2.

(iii) $\boldsymbol{v}^2$: $\bar{Y}_i$ is an unbiased estimator of $m(\Theta_i)$. Therefore a natural estimate of v^2 would be

$$\frac{1}{n-1}\sum_{i=1}^{n}(\bar{Y}_i-\hat{\mu})^2 \,.$$

However, this estimator is biased:

$$\begin{aligned}
\frac{1}{n-1}\sum_{i=1}^{n}\mathbb{E}[(\bar{Y}_i-\hat{\mu})^2] &= \frac{1}{n-1}\sum_{i=1}^{n}\mathbb{E}\left[\left(\frac{n-1}{n}\bar{Y}_i - \frac{1}{n}\sum_{j\neq i}\bar{Y}_j\right)^2\right] \\
&= \frac{n}{n-1}\mathbb{E}\left[\left(\frac{n-1}{n}\bar{Y}_1 - \frac{1}{n}\sum_{j=2}^{n}\bar{Y}_j\right)^2\right] \\
&= \frac{n}{n-1}\mathbb{E}\left[\left(\frac{n-1}{n}(\bar{Y}_1-\mu) - \frac{1}{n}\sum_{j=2}^{n}(\bar{Y}_j-\mu)\right)^2\right] \\
&= \frac{n}{n-1}\left(\left(\frac{n-1}{n}\right)^2\mathbb{E}[(\bar{Y}_1-\mu)^2] + \frac{n-1}{n^2}\mathbb{E}[(\bar{Y}_1-\mu)^2]\right) \\
&= \mathbb{E}\left[(\bar{Y}_1-\mu)^2\right] = \mathbb{E}\left[\frac{1}{m^2}\sum_{i=1}^{m}\sum_{j=1}^{m}Y_{1i}Y_{1j} - \frac{2}{m}\mu\sum_{j=1}^{m}Y_{1j} + \mu^2\right]
\end{aligned}$$

$$= \frac{1}{m}(\sigma^2 + v^2 + \mu^2) + \frac{m-1}{m}(v^2 + \mu^2) - \mu^2 = v^2 + \frac{\sigma^2}{m}.$$

We have to correct the estimator and get

$$\hat{v}^2 = \frac{1}{n-1}\sum_{i=1}^{n}(\bar{Y}_i - \hat{\mu})^2 - \frac{1}{m}\hat{\sigma}^2. \tag{3.6}$$

Example 3.1 A company has insured twenty similar collective risks over the last ten years. Table 3.2 shows the annual losses of each the risks. Thus the estimators are

$$\hat{\mu} = \frac{1}{20}\sum_{i=1}^{20} \bar{Y}_i = 102.02\,,$$

$$\hat{\sigma}^2 = \frac{1}{180}\sum_{i=1}^{20}\sum_{j=1}^{10}(Y_{ij} - \bar{Y}_i)^2 = 473.196$$

Table 3.2 Annual losses of the risks

i	1	2	3	4	5	6	7	8	9	10	$\bar{Y}_i$
1	67	71	50	56	64	69	77	94	56	48	65.2
2	80	82	109	61	89	113	149	91	127	108	100.9
3	125	118	101	135	120	117	101	101	100	122	114.0
4	96	144	152	124	94	132	155	143	94	113	124.7
5	176	161	153	191	139	157	192	151	175	147	164.2
6	89	129	95	88	131	72	91	122	79	56	95.2
7	70	116	102	129	81	69	75	120	95	93	95.0
8	22	33	48	20	56	25	43	53	48	64	41.2
9	121	106	55	103	87	81	130	113	98	97	99.1
10	126	101	158	129	110	112	107	114	117	95	116.9
11	125	106	104	135	83	177	157	95	101	128	121.1
12	116	134	133	167	142	150	134	153	134	109	137.2
13	53	62	89	76	63	66	69	70	57	49	65.4
14	110	136	75	52	49	85	110	89	68	34	80.8
15	125	112	145	114	61	74	168	98	114	97	110.8
16	87	95	85	111	89	101	85	88	90	164	99.5
17	92	46	74	82	68	94	116	109	81	89	85.1
18	44	74	67	93	89	57	83	53	82	38	68.0
19	95	107	147	154	121	125	146	83	117	139	123.4
20	103	105	139	138	145	145	140	148	137	127	132.7

Table 3.3 Credibility premia for the twenty risks

Risk	1	2	3	4	5	6	7
Premium	67.240	100.962	113.336	123.443	160.754	95.578	95.389
Risk	8	9	10	11	12	13	14
Premium	44.570	99.262	116.075	120.043	135.251	67.429	81.976
Risk	15	16	17	18	19	20	
Premium	110.313	99.640	86.038	69.885	122.215	131.000	

and

$$\hat{\upsilon}^2 = \frac{1}{19}\sum_{i=1}^{20}(\bar{Y}_i - \hat{\mu})^2 - \frac{1}{10}\hat{\sigma}^2 = 806.6\,.$$

The credibility factor can now be computed

$$Z = \frac{1}{1 + \dfrac{\hat{\sigma}^2}{\hat{\upsilon}^2 m}} = 0.944585\,.$$

Table 3.3 gives the credibility premia for the next year for each of the risks. The data had been simulated using $\mu = 100$, $\sigma^2 = 400$ and $\upsilon^2 = 900$. The credibility model interprets the data as having more fluctuations in a single risk than in the expected values of the risks. This is not surprising because we only have 10 data for each risk but 20 risks. Thus the fluctuations of $\bar{Y}_i$ could also be due to larger fluctuations within the risk. ■

3.3.2 The Bühlmann–Straub Model

The main fields of application for credibility theory are collective insurance contracts, e.g. employees insurance, fire insurance for companies, third party liability insurance for employees, travel insurance for the customers of a travel agent, etc. In reality the volume of the risk is not the same for each of the contracts. Looking at the credibility formulae calculated until now we can recognise that the credibility factors should be larger for higher risk volumes, because we have more data available. Thus we want to take the volume of the risk into consideration. The risk volume can be the number of employees, the sum insured, etc.

In order to see how to model different risk volumes we first consider an example.

Example 3.2 Assume that we are insuring taxi drivers. The losses of the drivers are dependent on each individual driver but also on the policies of the employer and the region where the company works. The i-th company employs P_i taxi drivers. There is a random variable Θ_i which determines the environment and the policies of the company. The random variable Θ_{ik} determines the risk class of each driver. We assume that the vectors $(\Theta_{i1}, \Theta_{i2}, \dots, \Theta_{iP_i}, \Theta_i)$ are independent and that, given Θ_i, the parameters $\Theta_{i1}, \Theta_{i2}, \dots, \Theta_{iP_i}$ are conditionally iid. The aggregate claims of different companies shall be independent. The aggregate claims of different drivers of company i are conditionally independent given Θ_i, the aggregate claims Y_{ikj} of a driver are conditionally independent given Θ_{ik} and Y_{ikj} depends on Θ_{ik} only. Then the expected value of the annual aggregate claims of company i given Θ_i is

$$\mathbb{E}\Bigl[\sum_{k=1}^{P_i} Y_{ikj} \Bigm| \Theta_i\Bigr] = P_i \mathbb{E}[Y_{i1j} \mid \Theta_i]\,.$$

The conditional variance of company i is then

$$\mathbb{E}\Bigl[\Bigl(\sum_{k=1}^{P_i} Y_{ikj} - P_i \mathbb{E}[Y_{i1j} \mid \Theta_i]\Bigr)^2 \Bigm| \Theta_i\Bigr] = P_i \mathbb{E}[(Y_{i1j} - \mathbb{E}[Y_{i1j} \mid \Theta_i])^2 \mid \Theta_i]\,.$$

Thus both the conditional expectation and the conditional variance are proportional to characteristics of the company. ■

The example will motivate the model assumptions. Let P_{ij} denote the volume of the i-th risk in year j. We assume that there exists a parameter Θ_i for each risk. The parameters $\Theta_1, \Theta_2, \dots, \Theta_n$ are assumed to be iid. Denote by Y_{ij} the aggregate claims of risk i in year j. We assume that the vectors $(Y_{i1}, Y_{i2}, \dots, Y_{im}, \Theta_i)$ are independent. Within each risk the random variables $Y_{i1}, Y_{i2}, \dots, Y_{im}$ are conditionally independent given Θ_i. Moreover, there exist functions $m(\vartheta)$ and $s^2(\vartheta)$ such that

$$\mathbb{E}[Y_{ij} \mid \Theta_i = \vartheta] = P_{ij} m(\vartheta)$$

and

$$\operatorname{Var}[Y_{ij} \mid \Theta_i = \vartheta] = P_{ij} s^2(\vartheta)\,.$$

As in the Bühlmann model we let $\mu = \mathbb{E}[m(\Theta_i)]$, $\sigma^2 = \mathbb{E}[s^2(\Theta_i)]$ and $v^2 = \operatorname{Var}[m(\Theta_i)]$. Denote by

$$P_{i\cdot} = \sum_{j=1}^{m} P_{ij}\,, \qquad P_{\cdot\cdot} = \sum_{i=1}^{n} P_{i\cdot}$$

the risk volume of the i-th risk over all m years and the risk volume over all years and companies.

First we normalise the annual losses. Let $X_{ij} = Y_{ij}/P_{ij}$. Note that $\mathbb{E}[X_{ij} \mid \Theta_i] = m(\Theta_i)$ and $\operatorname{Var}[X_{ij} \mid \Theta_i] = s^2(\Theta_i)/P_{ij}$. We next try to find the best linear estimator for $m(\Theta_i)$. Hence we have to minimise

$$\mathbb{E}\left[\left(a_{i0}+\sum_{j=1}^{m}a_{ij}X_{ij}-m(\Theta_i)\right)^2\right].$$

It is clear that X_{kj} does not carry information on $m(\Theta_i)$ if $k \neq i$. We first differentiate with respect to a_{i0},

$$\begin{aligned}\frac{1}{2}\frac{\mathrm{d}}{\mathrm{d}a_{i0}}\mathbb{E}\left[\left(a_{i0}+\sum_{j=1}^{m}a_{ij}X_{ij}-m(\Theta_i)\right)^2\right] &= \mathbb{E}\left[a_{i0}+\sum_{j=1}^{m}a_{ij}X_{ij}-m(\Theta_i)\right]\\ &= a_{i0}-\left(1-\sum_{j=1}^{m}a_{ij}\right)\mu = 0\,.\end{aligned}$$

As in the Bühlmann model we get

$$a_{i0}=\left(1-\sum_{j=1}^{m}a_{ij}\right)\mu$$

and therefore

$$\mathbb{E}\left[a_{i0}+\sum_{j=1}^{m}a_{ij}X_{ij}\right]=\mu\,.$$

Differentiate with respect to a_{ik},

$$\begin{aligned}0 &= \frac{1}{2}\frac{\mathrm{d}}{\mathrm{d}a_{ik}}\mathbb{E}\left[\left(a_{i0}+\sum_{j=1}^{m}a_{ij}X_{ij}-m(\Theta_i)\right)^2\right]\\ &= \mathbb{E}\left[X_{ik}\left(a_{i0}+\sum_{j=1}^{m}a_{ij}X_{ij}-m(\Theta_i)\right)\right]\\ &= a_{i0}\mu+\sum_{\substack{j=1\\ j\neq k}}^{m}a_{ij}\mathbb{E}[X_{ik}X_{ij}]+a_{ik}\mathbb{E}\left[X_{ik}^2\right]-\mathbb{E}[X_{ik}m(\Theta_i)]\,.\end{aligned}$$

Compute the terms. For $j \neq k$

$$\mathbb{E}[X_{ik}X_{ij}]=\mathbb{E}[\mathbb{E}[X_{ik}X_{ij}\mid\Theta_i]]=\mathbb{E}\left[(m(\Theta_i))^2\right]=v^2+\mu^2\,,$$

$$\mathbb{E}\left[X_{ik}^2\right]=\mathbb{E}\left[\mathbb{E}\left[X_{ik}^2\,\middle|\,\Theta_i\right]\right]=\mathbb{E}\left[(m(\Theta_i))^2+P_{ik}^{-1}s^2(\Theta_i)\right]=v^2+\mu^2+P_{ik}^{-1}\sigma^2\,,$$

$$\mathbb{E}[X_{ik}m(\Theta_i)]=\mathbb{E}[\mathbb{E}[X_{ik}m(\Theta_i)\mid\Theta_i]]=\mathbb{E}\left[(m(\Theta_i))^2\right]=v^2+\mu^2\,.$$

Thus

$$\left(1-\sum_{j=1}^{m} a_{ij}\right)\mu^2 + \sum_{\substack{j=1\\ j\neq k}}^{m} a_{ij}(v^2+\mu^2) + a_{ik}(v^2+\mu^2+P_{ik}^{-1}\sigma^2) - (v^2+\mu^2)$$

$$= a_{ik}P_{ik}^{-1}\sigma^2 - \left(1-\sum_{j=1}^{m} a_{ij}\right)v^2 = 0\,.$$

The right-hand side of

$$\sigma^2 P_{ik}^{-1} a_{ik} = \left(1-\sum_{j=1}^{m} a_{ij}\right)v^2$$

is independent of k and thus there exists a constant a such that $a_{ik} = P_{ik}a$. Then it follows readily that

$$a_{ik} = P_{ik}\frac{v^2}{\sigma^2 + P_{i\cdot}v^2}\,.$$

Note that the formula is consistent with (3.2). Define

$$\bar{X}_i = \frac{1}{P_{i\cdot}}\sum_{j=1}^{m} P_{ij}X_{ij}\,,$$

the weighted mean of the annual losses of the i-th risk. Then the credibility premium is

$$\left(1-\frac{P_{i\cdot}v^2}{\sigma^2+P_{i\cdot}v^2}\right)\mu + \frac{P_{i\cdot}v^2}{\sigma^2+P_{i\cdot}v^2}\bar{X}_i = Z_i\bar{X}_i + (1-Z_i)\mu$$

with the credibility factor

$$Z_i = \frac{1}{1+\dfrac{\sigma^2}{P_{i\cdot}v^2}}\,.$$

Note that the premium is consistent with (3.3). Denote by $\hat{P}_{i(m+1)}$ the (estimated) risk volume of the next year. Then the credibility premium for the next year will be

$$\hat{P}_{i(m+1)}(Z_i\bar{X}_i + (1-Z_i)\mu)\,.$$

The credibility factor Z_i depends on $P_{i\cdot}$. This means that in general different risks will get different credibility factors.

It remains to find estimators for the parameters.

(i) $\boldsymbol{\mu}$: We want to modify the estimator (3.4) for the Bühlmann–Straub model. The loss X_{ij} should be weighted with P_{ij} because it originates from a volume of this size. Thus we get

$$\hat{\mu} = \frac{1}{P_{\cdot\cdot}} \sum_{i=1}^{n} \sum_{j=1}^{m} P_{ij} X_{ij} = \frac{1}{P_{\cdot\cdot}} \sum_{i=1}^{n} P_{i\cdot} \bar{X}_i \,.$$

It turns out that $\hat{\mu}$ is unbiased.

(ii) $\boldsymbol{\sigma}^2$: We want to modify the estimator (3.5) for the Bühlmann–Straub model. The conditional variance of X_{ij} is $P_{ij}^{-1} s^2(\Theta_i)$. This suggests that we should weight the term $(X_{ij} - \bar{X}_i)^2$ with P_{ij}. Therefore we try an estimator of the form

$$c \sum_{i=1}^{n} \sum_{j=1}^{m} P_{ij} (X_{ij} - \bar{X}_i)^2 \,.$$

In order to compute the expected value of this estimator we keep i fixed. We have

$$\begin{aligned}
\mathbb{E}\Bigl[\sum_{j=1}^{m} P_{ij}(X_{ij} - \bar{X}_i)^2\Bigr] &= \mathbb{E}\Bigl[\sum_{j=1}^{m} P_{ij} X_{ij}^2 - 2 \sum_{j=1}^{m} P_{ij} X_{ij} \bar{X}_i + \sum_{j=1}^{m} P_{ij} \bar{X}_i^2\Bigr] \\
&= \mathbb{E}\Bigl[\sum_{j=1}^{m} P_{ij} X_{ij}^2\Bigr] - \mathbb{E}[P_{i\cdot} \bar{X}_i^2] \\
&= \sum_{j=1}^{m} P_{ij} \Bigl(\frac{\sigma^2}{P_{ij}} + v^2 + \mu^2\Bigr) - \frac{1}{P_{i\cdot}} \sum_{j=1}^{m} \sum_{k=1}^{m} P_{ij} P_{ik} \mathbb{E}[X_{ij} X_{ik}] \\
&= m\sigma^2 + P_{i\cdot}(v^2 + \mu^2) - P_{i\cdot}(v^2 + \mu^2) - \sigma^2 = (m-1)\sigma^2 \,. \qquad (3.7)
\end{aligned}$$

Hence the proposed estimator has expected value

$$\mathbb{E}\bigl[c \sum_{i=1}^{n} \sum_{j=1}^{m} P_{ij}(X_{ij} - \bar{X}_i)^2\bigr] = cn(m-1)\sigma^2 \,.$$

Thus our unbiased estimator is

$$\hat{\sigma}^2 = \frac{1}{n(m-1)} \sum_{i=1}^{n} \sum_{j=1}^{m} P_{ij}(X_{ij} - \bar{X}_i)^2 \,.$$

(iii) $\boldsymbol{v}^2$: A closer look at the estimator (3.6) shows that it contains the terms $(Y_{ij} - \hat{\mu})^2$, all with the same weight. For the estimator in the Bühlmann–Straub model we propose an estimator that contains a term of the form

$$\sum_{i=1}^{n} \sum_{j=1}^{m} P_{ij}(X_{ij} - \hat{\mu})^2 \,.$$

Let us compute its expected value,

$$
\begin{aligned}
\mathbb{E}\Big[\sum_{i=1}^{n}\sum_{j=1}^{m} P_{ij}(X_{ij}-\hat{\mu})^2\Big] &= \sum_{i=1}^{n}\sum_{j=1}^{m} P_{ij}\mathbb{E}[X_{ij}^2] - P_{..}\mathbb{E}[\hat{\mu}^2] \\
&= \sum_{i=1}^{n}\sum_{j=1}^{m} P_{ij}\Big(\frac{\sigma^2}{P_{ij}} + v^2 + \mu^2\Big) \\
&\quad - \frac{1}{P_{..}}\sum_{i=1}^{n}\sum_{j=1}^{m}\sum_{k=1}^{n}\sum_{l=1}^{m} P_{ij}P_{kl}\mathbb{E}[X_{ij}X_{kl}] \\
&= nm\sigma^2 + P_{..}(v^2+\mu^2) - \frac{1}{P_{..}}\Big(\sum_{i=1}^{n}\sum_{j=1}^{m} P_{ij}^2\Big(\frac{\sigma^2}{P_{ij}} + v^2 + \mu^2\Big) \\
&\quad + \sum_{i=1}^{n}\sum_{j=1}^{m}\sum_{\substack{l=1\\ l\neq j}}^{m} P_{ij}P_{il}(v^2+\mu^2) + \sum_{i=1}^{n}\sum_{j=1}^{m}\sum_{\substack{k=1\\ k\neq i}}^{n}\sum_{l=1}^{m} P_{ij}P_{kl}\mu^2\Big) \\
&= (nm-1)\sigma^2 + \Big(P_{..} - \sum_{i=1}^{n}\frac{P_{i.}^2}{P_{..}}\Big)v^2 . \qquad (3.8)
\end{aligned}
$$

Let

$$
P^* = \frac{1}{nm-1}\sum_{i=1}^{n} P_{i.}\Big(1 - \frac{P_{i.}}{P_{..}}\Big) .
$$

Then

$$
\hat{v}^2 = \frac{1}{P^*}\Big\{\Big(\frac{1}{nm-1}\sum_{i=1}^{n}\sum_{j=1}^{m} P_{ij}(X_{ij}-\hat{\mu})^2\Big) - \hat{\sigma}^2\Big\}
$$

is an unbiased estimator of v^2.

Example 3.3 An insurance company has issued motor insurance policies for business cars to six companies. Table 3.4 shows the number of cars for each of the six companies in each of the past ten years. Table 3.5 shows the aggregate claims, measured in thousands of €, for each of the companies in each of the past ten years. The first step is to calculate the mean aggregate claims per car $X_{ij} = Y_{ij}/P_{ij}$ for each of the companies in each year, see Table 3.6. For the computation of the estimators the figures in Table 3.7 are needed. We further get

$$
P_{..} = 554, \qquad P^* = 7.08585 .
$$

We can now compute the following estimates.

Table 3.4 Premium volumes

P_{ij}	Year										
		1	2	3	4	5	6	7	8	9	10
Company number	1	10	10	12	12	15	15	15	13	9	9
	2	5	5	5	5	5	5	5	5	5	5
	3	2	2	3	3	3	3	3	3	3	3
	4	15	20	20	20	21	22	22	22	25	25
	5	5	5	5	3	3	3	3	3	3	5
	6	10	10	10	10	10	10	10	12	12	12

Table 3.5 Aggregate annual losses

Y_{ij}	Year										
		1	2	3	4	5	6	7	8	9	10
Company number	1	74	50	180	43	179	140	95	149	20	81
	2	52	111	83	74	87	85	40	71	93	83
	3	40	28	60	59	43	74	61	46	72	81
	4	171	85	100	116	153	44	13	31	110	252
	5	49	132	74	37	21	54	27	43	53	102
	6	126	148	128	151	165	128	100	65	233	246

$$\hat{\mu} = \frac{1}{P_{\cdot\cdot}} \sum_{i=1}^{6} P_{i\cdot} \bar{X}_i = 9.94765\,,$$

$$\hat{\sigma}^2 = \frac{1}{6} \sum_{i=1}^{6} \frac{1}{9} \sum_{j=1}^{10} P_{ij} (X_{ij} - \bar{X}_i)^2 = 157.808\,,$$

$$\hat{v}^2 = \frac{1}{P*} \left[\frac{1}{59} \sum_{i=1}^{6} \sum_{j=1}^{10} P_{ij} (X_{ij} - \hat{\mu})^2 - \hat{\sigma}^2 \right] = 28.7578\,.$$

The number of business cars each company plans to use in the next year and the credibility premium is given in Table 3.8. One can clearly see how the credibility factor increases with $P_{i\cdot}$. ■

Table 3.6 Normalised losses

Year	1	2	3	4	5	6	7	8	9	10
1	7.4	5	15	3.583	11.933	9.333	6.333	11.462	2.222	9
2	10.4	22.2	16.6	14.8	17.4	17	8	14.2	18.6	16.6
3	20	14	20	19.667	14.333	24.667	20.333	15.333	24	27
4	11.4	4.25	5	5.8	7.286	2	0.591	1.409	4.4	10.08
5	9.8	26.4	14.8	12.333	7	18	9	14.333	17.667	20.4
6	12.6	14.8	12.8	15.1	16.5	12.8	10	5.417	19.417	20.5

Table 3.7 Quantities used in the estimators

Company	$P_{i\cdot}$	$\bar{X}_i$	$\sum_j P_{ij}(X_{ij} - \bar{X}_i)^2$	$\sum_j P_{ij}(X_{ij} - \hat{\mu})^2$
1	120	8.425	1659.62	1937.84
2	50	15.580	735.78	2321.95
3	28	20.143	494.10	3404.48
4	212	5.071	2310.63	7352.86
5	38	15.579	1289.26	2494.30
6	106	14.057	2032.23	3821.88

Table 3.8 Premium next year

Company	Cars next year	Credibility factor	Credibility premium per unit volume	Actual premium
1	9	0.95627	8.4916	76.424
2	6	0.90110	15.0230	90.138
3	3	0.83613	18.4722	55.417
4	24	0.97477	5.1938	124.651
5	3	0.87382	14.8684	44.605
6	12	0.95078	13.8544	166.252

3.3.3 *The Bühlmann–Straub Model with Missing Data*

In practice it is not realistic that all policy holders of a certain collective insurance type started to insure their risks in the same year. Thus some of the P_{ij}'s may be 0. The question arises whether this has an influence on the model. The answer is that its influence is only small. Denote by m_i the number of years with $P_{ij} \neq 0$.

For the credibility estimator it is clear that $a_{ij} = 0$ if $P_{ij} = 0$ because then also $Y_{ij} = 0$. For convenience we define $X_{ij} = 0$ if $P_{ij} = 0$. The computation of a_{ij}

does not change if some of the P_{ij}'s are 0. Moreover, it is easy to see that $\hat{\mu}$ remains unbiased.

Next let us consider the estimator of σ^2. A closer look at (3.7) shows that

$$\mathbb{E}\Bigl[\sum_{j=1}^{m} P_{ij}(X_{ij} - \bar{X}_i)^2\Bigr] = (m_i - 1)\sigma^2 .$$

Hence the estimator of σ^2 must be changed to

$$\hat{\sigma}^2 = \frac{1}{n}\sum_{i=1}^{n} \frac{1}{m_i - 1}\sum_{j=1}^{m} P_{ij}(X_{ij} - \bar{X}_i)^2 .$$

Consider the estimator of v. The computation (3.8) changes to

$$\mathbb{E}\Bigl[\sum_{i=1}^{n}\sum_{j=1}^{m} P_{ij}(X_{ij} - \hat{\mu})^2\Bigr] = (m_{\cdot} - 1)\sigma^2 + \Bigl(P_{\cdot\cdot} - \sum_{i=1}^{n} \frac{P_{i\cdot}^2}{P_{\cdot\cdot}}\Bigr)v^2 ,$$

where

$$m_{\cdot} = \sum_{i=1}^{n} m_i .$$

Thus one has to change

$$P^* = \frac{1}{m_{\cdot} - 1}\sum_{i=1}^{n} P_{i\cdot}\Bigl(1 - \frac{P_{i\cdot}}{P_{\cdot\cdot}}\Bigr)$$

and

$$\hat{v}^2 = \frac{1}{P^*}\Bigl\{\Bigl(\frac{1}{m_{\cdot} - 1}\sum_{i=1}^{n}\sum_{j=1}^{m} P_{ij}(X_{ij} - \hat{\mu})^2\Bigr) - \hat{\sigma}^2\Bigr\}.$$

3.4 General Bayes Methods

We can generalise the approach above by summarising the observations in a vector $\boldsymbol{X}_i$. In the models considered before we would have $\boldsymbol{X}_i = (X_{i,1}, X_{i,2}, \dots, X_{i,m_i})^\top$. We then want to estimate the vector $\boldsymbol{m}_i \in \mathbb{R}^s$. In the models above we had $\boldsymbol{m}_i = m(\Theta_i)$. We assume that the vectors $(\boldsymbol{X}_i^\top, \boldsymbol{m}_i^\top)^\top$ are independent and that second moments exist. In contrast to the models considered above we do not assume that the X_{ik} are conditionally independent or identically distributed. For example, we could have

$$\boldsymbol{m}_i = (\mathbb{E}[X_{ik} \mid \Theta_i], \mathbb{E}[X_{ik}^2 \mid \Theta_i])^\top$$

and

$$X_i = (X_{i1}, X_{i1}^2, X_{i2}, X_{i2}^2, \ldots, X_{i,m_i}, X_{i,m_i}^2)^\top .$$

The goal is now to estimate $\boldsymbol{m}_i$.

The estimator $\boldsymbol{M}$ minimising $\mathbb{E}[\|\boldsymbol{M} - \boldsymbol{m}_i\|^2]$ is then $\boldsymbol{M} = \mathbb{E}[\boldsymbol{m}_i \mid \boldsymbol{X}_i]$. The problem with the estimator is again that we need the joint distribution of the quantities. We therefore restrict again to linear estimators, i.e. estimators of the form $\boldsymbol{M} = \boldsymbol{g} + \boldsymbol{G}\boldsymbol{X}$, where, for the moment, we omit the index. We denote the entries of $\boldsymbol{g}$ by (g_{i0}). The quantity to minimise is then

$$\rho(\boldsymbol{M}) := \sum_i \mathbb{E}\Bigl[\Bigl(g_{i0} + \sum_{j=1}^m g_{ij} X_j - m_i\Bigr)^2\Bigr].$$

In order to minimise $\rho(\boldsymbol{M})$ we first take the derivative with respect to g_{i0} and equate it to zero

$$\frac{\mathrm{d}\rho(\boldsymbol{M})}{\mathrm{d}g_{i0}} = 2\mathbb{E}\Bigl[g_{i0} + \sum_{\ell=1}^m g_{i\ell} X_\ell - m_i\Bigr] = 0\,.$$

The derivative with respect to g_{ij} yields

$$\frac{\mathrm{d}\rho(\boldsymbol{M})}{\mathrm{d}g_{ij}} = 2\mathbb{E}\Bigl[X_j\Bigl(g_{i0} + \sum_{\ell=1}^m g_{i\ell} X_\ell - m_i\Bigr)\Bigr] = 0\,.$$

In matrix form we can express the equations as

$$\begin{aligned}\mathbb{E}[\boldsymbol{g} + \boldsymbol{G}\boldsymbol{X} - \boldsymbol{m}] &= \boldsymbol{0}\,,\\ \mathbb{E}[(\boldsymbol{g} + \boldsymbol{G}\boldsymbol{X} - \boldsymbol{m})\boldsymbol{X}^\top] &= \boldsymbol{0}\,.\end{aligned}$$

The solution we are looking for is therefore

$$\boldsymbol{g} = \mathbb{E}[\boldsymbol{m}] - \boldsymbol{G}\mathbb{E}[\boldsymbol{X}]\,, \tag{3.9a}$$

$$\boldsymbol{G} = \mathrm{Cov}[\boldsymbol{m}, \boldsymbol{X}^\top]\mathrm{Var}[\boldsymbol{X}]^{-1}\,. \tag{3.9b}$$

Here we assume that $\mathrm{Var}[\boldsymbol{X}]$ is invertible. Note that $\mathrm{Var}[\boldsymbol{X}]$ is symmetric. Because for any vector $\boldsymbol{a}$ we have $\boldsymbol{a}^\top\mathrm{Var}[\boldsymbol{X}]\boldsymbol{a} = \mathrm{Var}[\boldsymbol{a}^\top\boldsymbol{X}]$ it follows that $\mathrm{Var}[\boldsymbol{X}]$ is positive semi-definite. If $\mathrm{Var}[\boldsymbol{X}]$ is not invertible, then there is an $\boldsymbol{a}$ such that $\boldsymbol{a}^\top\mathrm{Var}[\boldsymbol{X}]\boldsymbol{a} = 0$, or equivalently, $\boldsymbol{a}^\top\boldsymbol{X} = 0$. Hence some of the entries of $\boldsymbol{X}$ can be expressed via the others. In this case, there is no loss of generality in reducing $\boldsymbol{X}$ to $\boldsymbol{X}^*$ such that $\mathrm{Var}[\boldsymbol{X}^*]$ becomes invertible.

If we now again consider n collective contracts $(\boldsymbol{X}_i, \boldsymbol{m}_i)$ the *linear Bayes estimator* $\boldsymbol{M}_i$ for $\boldsymbol{m}_i$ can be written as

$$M_i = \mathbb{E}[m_i] + \mathrm{Cov}[m_i, X_i^\top]\mathrm{Var}[X_i]^{-1}(X_i - \mathbb{E}[X_i]) .$$

Thus we start with the expected value $\mathbb{E}[m_i]$ and correct it according to the deviation of X_i from the expected value $\mathbb{E}[X_i]$. The correction is "larger" if the covariance between m_i and X_i is "larger" and "smaller" if the variance of X_i is "larger".

Example 3.4 Suppose there is an unobserved variable Θ_i that determines the risk. Suppose $\mathbb{E}[X_i \mid \Theta_i] = Y_i b(\Theta_i)$ and $\mathrm{Var}[X_i \mid \Theta_i] = P_i V_i(\Theta_i)$ for some known $Y_i \in \mathbb{R}^{m\times q}$, $P_i \in \mathbb{R}^{m\times m}$ and some functions $b(\theta) \in \mathbb{R}^q$, $V_i(\theta) \in \mathbb{R}^{m\times m}$ of the unobservable random variable Θ_i. We assume that P_i is invertible. The Bühlmann and the Bühlmann–Straub models are special cases.

Here we are interested in estimating $b(\Theta_i)$. Introduce the following quantities $\beta = \mathbb{E}[b(\Theta_i)]$, $\Lambda = \mathrm{Var}[b(\Theta_i)]$ and $\Phi_i = P_i \mathbb{E}[V_i(\Theta_i)]$. The moments required are β, $\mathbb{E}[X_i] = Y_i\beta$,

$$\begin{aligned}\mathrm{Cov}[b(\Theta_i), X_i^\top] &= \mathrm{Cov}[\mathbb{E}[b(\Theta_i) \mid \Theta_i], \mathbb{E}[X_i^\top \mid \Theta_i]] + \mathbb{E}[\mathrm{Cov}[b(\Theta_i), X_i^\top \mid \Theta_i]] \\ &= \mathrm{Cov}[b(\Theta_i), b(\Theta_i)^\top Y_i^\top] = \mathrm{Cov}[b(\Theta_i), b(\Theta_i)^\top]Y_i^\top = \Lambda Y_i^\top ,\end{aligned}$$

and

$$\mathrm{Var}[X_i] = \mathrm{Var}[\mathbb{E}[X_i \mid \Theta_i]] + \mathbb{E}[\mathrm{Var}[X_i \mid \Theta_i]] = Y_i \Lambda Y_i^\top + \Phi_i .$$

We assume that $\mathbb{E}[V(\Theta_i)]$ and therefore Φ_i is invertible. This yields the solution

$$\bar{b}_i = \Lambda Y_i^\top (Y_i \Lambda Y_i^\top + \Phi_i)^{-1} X_i + (I - \Lambda Y_i^\top (Y_i \Lambda Y_i^\top + \Phi_i)^{-1} Y_i)\beta . \tag{3.10}$$

Note that Λ is symmetric. If Λ were not invertible, we could write some coordinates of $b(\Theta_i)$ as a linear function of the others. Thus we assume that Λ is invertible. A well-known formula gives

$$(Y_i \Lambda Y_i^\top + \Phi_i)^{-1} = \Phi_i^{-1} - \Phi_i^{-1} Y_i (\Lambda^{-1} + Y_i^\top \Phi_i^{-1} Y_i)^{-1} Y_i^\top \Phi_i^{-1} .$$

Define the matrix

$$\begin{aligned}Z_i &= \Lambda Y_i^\top \Phi_i^{-1} Y_i (\Lambda^{-1} + Y_i^\top \Phi_i^{-1} Y_i)^{-1} \Lambda^{-1} \\ &= \Lambda Y_i^\top \Phi_i^{-1} Y_i (\Lambda Y_i^\top \Phi_i^{-1} Y_i + I)^{-1} .\end{aligned}$$

Then we find

$$\begin{aligned}Z_i &= (\Lambda Y_i^\top \Phi_i^{-1} Y_i + I - I)(\Lambda Y_i^\top \Phi_i^{-1} Y_i + I)^{-1} \\ &= I - (\Lambda Y_i^\top \Phi_i^{-1} Y_i + I)^{-1} .\end{aligned}$$

We also find

$$
\begin{aligned}
&\Lambda Y_i^\top (Y_i \Lambda Y_i^\top + \Phi_i)^{-1} Y_i \\
&\quad = \Lambda Y_i^\top \Phi_i^{-1} Y_i - \Lambda Y_i^\top \Phi_i^{-1} Y_i (\Lambda^{-1} + Y_i^\top \Phi_i^{-1} Y_i)^{-1} Y_i^\top \Phi_i^{-1} Y_i \\
&\quad = (I - Z_i) \Lambda Y_i^\top \Phi_i^{-1} Y_i = (\Lambda Y_i^\top \Phi_i^{-1} Y_i + I)^{-1} \Lambda Y_i^\top \Phi_i^{-1} Y_i = Z_i \,.
\end{aligned}
$$

It follows that we can write the LB estimator as

$$
\bar{b}_i = (I - Z_i)(\Lambda Y_i^\top \Phi_i^{-1} X_i + \beta) \,. \tag{3.11}
$$

The matrix Z_i is called the *credibility matrix.*

Suppose that in addition $Y_i^\top \Phi_i^{-1} Y_i$ is invertible. This is equivalent to Y_i having full rank q and $m \geq q$. Then

$$
(\Lambda Y_i^\top \Phi_i^{-1} Y_i + I)(Y_i^\top \Phi_i^{-1} Y_i)^{-1} Y_i^\top \Phi_i^{-1} = (\Lambda + (Y_i^\top \Phi_i^{-1} Y_i)^{-1}) Y_i^\top \Phi_i^{-1} \,.
$$

Multiplying by $(\Lambda Y_i^\top \Phi_i^{-1} Y_i + I)^{-1}$ on the left gives

$$
(Y_i^\top \Phi_i^{-1} Y_i)^{-1} Y_i^\top \Phi_i^{-1} = (\Lambda Y_i^\top \Phi_i^{-1} Y_i + I)^{-1} (\Lambda + (Y_i^\top \Phi_i^{-1} Y_i)^{-1}) Y_i^\top \Phi_i^{-1} \,.
$$

Rearranging the terms yields

$$
\begin{aligned}
(I - Z_i) \Lambda Y_i^\top \Phi_i^{-1} &= (\Lambda Y_i^\top \Phi_i^{-1} Y_i + I)^{-1} \Lambda Y_i^\top \Phi_i^{-1} \\
&= (I - (\Lambda Y_i^\top \Phi_i^{-1} Y_i + I)^{-1})(Y_i^\top \Phi_i^{-1} Y_i)^{-1} Y_i^\top \Phi_i^{-1} \\
&= Z_i (Y_i^\top \Phi_i^{-1} Y_i)^{-1} Y_i^\top \Phi_i^{-1} \,.
\end{aligned}
$$

If we define

$$
\hat{b}_i = (Y_i^\top \Phi_i^{-1} Y_i)^{-1} Y_i^\top \Phi_i^{-1} X_i
$$

we can express the estimator in credibility weighted form

$$
\bar{b}_i = Z_i \hat{b}_i + (I - Z_i) \beta \,.
$$

■

3.5 Hilbert Space Methods

In the previous section we considered the space of quadratic integrable random variables $\mathcal{L}^2$. Choose the inner product $\langle X, Y \rangle = \mathbb{E}[X^\top Y]$. The problem was to minimise $\|M - m\|$, where M was an estimator from some linear subspace of estimators. We now want to consider the problem in the Hilbert space $(\mathcal{L}, \langle \cdot, \cdot \rangle)$.

Let $m \in \mathcal{L}^2$ be an unknown quantity. Let $\mathcal{L} \subset \mathcal{L}^2$ be some subset. We now want to minimise $\rho(M) = \|M - m\|^2$ over all estimators $M \in \mathcal{L}$. If $\mathcal{L}$ is a closed subspace of $\mathcal{L}^2$ then the optimal estimator, called the *$\mathcal{L}$-Bayes estimator*, is the projection

$\boldsymbol{m}_L = \text{pro}(\boldsymbol{m} \mid \mathcal{L})$. Because $\boldsymbol{m} - \boldsymbol{m}_{\mathcal{L}} \in \mathcal{L}^{\perp}$ we have $\rho(\boldsymbol{M}) = \|\boldsymbol{m} - \boldsymbol{m}_{\mathcal{L}}\|^2 + \|\boldsymbol{m}_{\mathcal{L}} - \boldsymbol{M}\|^2$ which clearly is minimised by choosing $\boldsymbol{M} = \boldsymbol{m}_{\mathcal{L}}$.

If $\mathcal{L}' \subset \mathcal{L}$ is a closed linear subspace then iterated projections give $\boldsymbol{m}_{\mathcal{L}'} = \text{pro}(\boldsymbol{m}_{\mathcal{L}} \mid \mathcal{L}')$. More generally, if $\{\boldsymbol{0}\} = \mathcal{L}_0 \subset \mathcal{L}_1 \subset \cdots \subset \mathcal{L}_n = \mathcal{L}^2$ is a nested family of closed linear subspaces of $\mathcal{L}^2$. Then we have $\boldsymbol{m} = \boldsymbol{m}_{\mathcal{L}_n}$ and

$$\boldsymbol{m}_{\mathcal{L}_k} = \sum_{j=1}^{k} (\boldsymbol{m}_{\mathcal{L}_j} - \boldsymbol{m}_{\mathcal{L}_{j-1}}) \,.$$

For example, if $\mathcal{L}_1$ is the space of constants and $\mathcal{L}_2$ is the space of linear estimators we find $\boldsymbol{m}_{\mathcal{L}_1} = \mathbb{E}[\boldsymbol{m}_i]$ and

$$\boldsymbol{m}_{\mathcal{L}_2} - \boldsymbol{m}_{\mathcal{L}_1} = \text{Cov}[\boldsymbol{m}_i, \boldsymbol{X}_I^\top]\text{Var}[\boldsymbol{X}_i]^{-1}(\boldsymbol{X}_i - \mathbb{E}[\boldsymbol{X}_i]) \,.$$

3.6 Bonus-Malus Systems

Credibility is a method that works well for collective contracts. For individual customers, the method is not feasible because the customers will not understand the premium policy. One therefore looks for a scheme that is easier to understand. For motor insurance, a traditional method is to use bonus-malus systems.

A bonus-malus system consists of I classes $\{1, 2, \dots, I\}$, a transition rule $\boldsymbol{T}(k) = (t_{ij}(k))_{i,j}$, a premium scale $\boldsymbol{b}$, an initial state i_0, and a premium π for the maximal malus I. Here, $t_{ij}(k) = 1$ if a customer in class i is transferred to class j if he has k claims, and $t_{ij}(k) = 0$ otherwise. In particular, $\sum_j t_{ij}(k) = 1$. The premium scale is $\boldsymbol{b} = (b_1, b_2, \dots, b_I)^\top$ with $0 < b_1 \le b_2 \le \cdots \le b_I = 1$.

A customer now has N_i claims in year i. Let $X(0)$ be the unit row vector with $X_i(0) = \mathbb{I}_{i=i_0}$. Then the process $X(n) = X(n-1)\boldsymbol{T}(N_n)$ models the movement through the classes of the customer. The premium the customer has to pay in period n is then $X(n-1)\boldsymbol{b}\pi$.

Suppose now that $\{N_i\}$ are iid. Then $\{X(n)\}$ is a Markov chain with transition probabilities $\boldsymbol{P} = \sum_{k=0}^{\infty} \boldsymbol{T}(k)\mathbb{P}[N_n = k]$. This Markov chain is ergodic. There is a stationary distribution $\boldsymbol{p} = (p_1, p_2, \dots, p_I)$. Thus, the average premium earned per year converges to $\boldsymbol{p}\boldsymbol{b}\pi$. The net premium the insurer should charge is therefore $\pi = \mathbb{E}[N_k]\mu/(\boldsymbol{p}\boldsymbol{b})$, where μ is the expected claim size.

Of course, in such a scheme there are many customers. Suppose now there is a risk parameter Θ. Denote its distribution by $F_\theta(\vartheta)$. Then the stationary distribution $\boldsymbol{p}(\vartheta)$ depends on the risk parameter. The net premium is then $\int \mathbb{E}[N_k \mid \Theta = \vartheta]\mu(\vartheta)/(\boldsymbol{p}(\vartheta)\boldsymbol{b}) \,\mathrm{d}F_\theta(\vartheta)$.

In a credibility system, the ultimate premium charged to a customer with risk parameter ϑ is $\mathbb{E}[N_k \mid \Theta = \vartheta]\mu(\vartheta)/(\boldsymbol{p}(\vartheta)\boldsymbol{b})$. In a bonus-malus system one would like to charge this premium. However, because only a finite number of premia are possible, this may not be possible in any case. One possible criterion could be to minimise the square loss

$$\int [\boldsymbol{p}(\vartheta)\boldsymbol{b}\pi - \mathbb{E}[N_k \mid \Theta = \vartheta]\mu(\vartheta)]^2 \, \mathrm{d}F_\theta(\vartheta) \, .$$

This problem of finding the optimal $\boldsymbol{b}$ is quite complicated because the premium π depends on $\boldsymbol{b}$ in a non-linear way.

Bibliographical Remarks

Credibility theory originated in North America in the early part of the twentieth century. These early ideas are now known as *American credibility*. A first reference is [137]. References can also be found in the survey paper by Norberg [103]. The Bayesian approach to credibility is generally attributed to Bailey [16, 17]. The empirical Bayes approach is due to Bühlmann [27] and Bühlmann and Straub [31]. Hilbert space methods were considered in [67]. An early reference to bonus-malus systems is [79]. An introduction to bonus-malus systems can be found in [95] or [110, Chap. 7] and references therein. Further literature on the topic may be found, for example, in [65, 70, 78, 88, 91].

Chapter 4
Claims Reserving

We review the classical methods for claims reserving and we further consider a mathematically interesting model that is quite simply constructed.

4.1 Introduction

In insurance, premia are usually paid annually. The net premia are an a priori prediction for claims occurring in the year the contract covers. At the end of the year, more information on claims is known. Some of the claims may be settled, but not all of them. The following knowledge on claims is possible:

- Claims are completely settled. For such claims nothing has to be done.
- It is known that a claim has occurred, but it is not (completely) settled. For such claims, additional payments may be necessary in the future. Thus, reserves have to be built on the balance sheet in order to have the money for the settlement. This type of claim is called *RBNS*, Reported But Not Settled.
- Claims have occurred, but the insurer does not know. For such claims, reserves have to be built. This type of claim is called *IBNYR*, Incurred But Not Yet Reported.

A typical example could be health insurance. If a customer sees a doctor, he often waits until he gets the invoice before informing his health insurer. Then the insurer may not know about the claim at the end of the year. Or after an accident, the insurer has to pay salary replacements. The insurer knows that the claim has happened but cannot know for how long these salary replacements have to be paid for.

In order to build the reserves, the insurer has to predict the liabilities from past events. This can be done in the following way:

- One builds a reserve for each of the not yet settled claims and a reserve for the IBNYR claims. This would cause a lot of work because an estimate for each of the RBNS claims has to be obtained.

H. Schmidli, *Risk Theory*, Springer Actuarial,
DOI:10.1007/978-3-319-72005-0_4

Table 4.1 Run-off triangle

Occurrence year	Development year			
	0	1	2	3
1	$X_{1,0}$	$X_{1,1}$	$X_{1,2}$	$X_{1,3}$
2	$X_{2,0}$	$X_{2,1}$	$X_{2,2}$	
3	$X_{3,0}$	$X_{3,1}$		
4	$X_{4,0}$			

Occurrence year	Development year			
	0	1	2	3
1	$C_{1,0}$	$C_{1,1}$	$C_{1,2}$	$C_{1,3}$
2	$C_{2,0}$	$C_{2,1}$	$C_{2,2}$	
3	$C_{3,0}$	$C_{3,1}$		
4	$C_{4,0}$			

- One builds a reserve for the aggregate RBNS claims and one for the IBNYR claims.
- One builds a single reserve for all future liabilities.

Sometimes, a combination of the methods is used. For some special larger claims with more individual information and a non-typical claim development, a reserve is built and the other liabilities are treated together. For example, for someone injured in a car accident, one may have more detailed estimates based on the severity of the injuries.

The data available are usually represented in a *run-off triangle* (Table 4.1). Claims from the occurrence year $i = 1, 2, 3, \ldots$ are paid in the development year $j = 0, 1, 2, \ldots$. In the triangle, one can represent the payment $X_{i,j}$ of the claims of year i and development year j or the accumulated payments $C_{i,j} = \sum_{\ell=0}^{j} X_{i,\ell}$. Thus $X_{i,j}$ represents the payment in accounting year $i + j$ for the claims which occurred in year i.

It is important that the payments are comparable. That means the payments have to be adjusted to inflation. Usually, the present year is taken as a basis. If the present year is n, the payments $X_{n-k,k}$ for $k \le n$ are the real payments, the other payments need to be adjusted. The payments probably also need to be adjusted to changes in the regulations or court decisions. For example, suppose an insurer had the practice to pay at most the present value of a car damaged in a car accident. If a court then decided that the full costs for repairing the car have to be covered, even if the value of the car is less, then the past payments have to be adjusted. One may have to estimate the increase of the costs.

Additional information on the portfolio are the premia π_i collected. Another possible piece of information is the risk volume v_i. For example, in car insurance one may know the annual kilometres driven, or the present value of the cars insured. In the sequel we will denote by n the number of years one has information on. We further will assume that there exists a J such that $C_{i,J}$ denotes the aggregate claim size for the occurrence year i. This means that $X_{i,j} = 0$ for all $j > J$. In order to be able to make a prediction, we should assume $J < n$. Otherwise, external predictions are needed.

In order to calculate the reserves needed, we have to meet two competing tasks for the method

- The method should be justified by a well fitted mathematical model and be optimal with respect to some criteria.
- The method should be simple and easy to implement.

A good mathematical model will often lead to a complicated method that is not possible to implement. For a simple method the model risk becomes large. In practice, however, several methods have turned out to be easy to implement and have led to results that are not too poor.

4.2 Classical Claims Reserving Methods

4.2.1 The Chain-Ladder Method

One of the most popular methods in practice is the chain-ladder method. The method is very simple and thus easily practicable. We consider the accumulated payments $C_{i,j}$. The intuition behind the method is that there are chain-ladder factors φ_j such that a good prediction for next year's accumulated payments is $C_{i,j}\varphi_{j+1}$. In other words,

$$C_{i,j+1} = C_{i,j}(\varphi_{j+1} + \text{error}) \;.$$

This is a linear regression without intercept. It is also assumed that larger $C_{i,j}$ contain more information. This makes sense because a larger risk volume will lead to a larger accumulated claim and therefore also to larger accumulated claim payments. Therefore, a natural estimate for φ_j is

$$\hat{\varphi}_j = \frac{\sum_{i=1}^{n-j} C_{i,j}}{\sum_{i=1}^{n-j} C_{i,j-1}} \;.$$

A prediction for the accumulated payments in year $i + j$ for the occurrence year i with $i + j > n$ is then

$$\hat{C}_{i,j} = C_{i,n-i} \prod_{\ell=n-i+1}^{j} \hat{\varphi}_\ell \;.$$

If $J < n$, one can predict the reserves needed for the occurrence year i,

$$R_i = \hat{C}_{i,J} - C_{i,n-i} = C_{i,n-i}\Big(\prod_{j=n-i+1}^{J} \hat{\varphi}_j - 1\Big).$$

For some reserving methods considered later, we will need claims quotas. These are estimates of the part of the total claims that will be paid until development year j. From the chain-ladder factors, it is possible to get estimates for the claims quotas (if $J < n$)

$$\hat{\gamma}_j = \frac{\hat{C}_{i,J}}{\hat{C}_{i,j}} = \prod_{\ell=j+1}^{J} \frac{1}{\hat{\varphi}_\ell},$$

where we let $\hat{C}_{i,n-i} = C_{i,n-i}$.

4.2.2 The Loss-Development Method

It is assumed that

$$C_{i,j} = \gamma_j C_{i,J} + \text{error}$$

for some claims quotas γ_j. The claims quotas need to be given externally, or could be estimated by another claims reserving method such as, for example, the chain-ladder method. We can predict the total claims (if $n < J + i$)

$$\hat{C}_{i,J} = \frac{C_{i,n-i}}{\gamma_{n-i}}.$$

The reserves needed therefore are

$$R_i = C_{i,n-i}\Big(\frac{1}{\gamma_{n-i}} - 1\Big).$$

A prediction for the accumulated payments until development year $n-i \le j \le J$ is

$$\hat{C}_{i,j} = \frac{\gamma_j C_{i,n-i}}{\gamma_{n-i}}.$$

This is equivalent to the chain-ladder reserves if one chooses $\hat{\gamma}_j$ as estimates for γ_j.

4.2.3 The Additive Method

It is assumed that

$$X_{i,j} = v_i \zeta_j + \text{error}.$$

Instead of the risk volume v_i one could also use the premium π_i. With the premium π_i, the factors ζ_j have the interpretation that our predictions for the payments $X_{i,j}$ for the development year j is $\pi_i \zeta_j$. Note that the observed payments $\{X_{i,j} : j \leq n-i\}$ do not change the predictions $\hat{X}_{i,j}$ for $j > n-i$. We again suppose that there is more information in higher payments and risk volumes, respectively. Thus a natural estimate for the factor ζ_j is

$$\hat{\zeta}_j = \frac{\sum_{i=1}^{J-j} X_{i,j}}{\sum_{i=1}^{J-j} v_i} .$$

The prediction for the payments $X_{i,j}$ in development year $j > n-i$ then becomes

$$\hat{X}_{i,j} = \hat{\zeta}_j v_i .$$

Thus the reserve R_i for occurrence year i becomes (if $n > J$)

$$R_i = \sum_{j=n-i+1}^{J} \hat{X}_{i,j} = v_i \sum_{j=n-i+1}^{J} \hat{\zeta}_j .$$

The prediction for the total payments is

$$\hat{C}_{i,J} = C_{i,n-i} + R_i = C_{i,n-i} + v_i \sum_{j=n-i+1}^{J} \hat{\zeta}_j .$$

An estimate for the claim quota $\kappa = \sum_{j=0}^{J} \zeta_j$ with respect to the risk volume is

$$\hat{\kappa} = \sum_{j=0}^{J} \hat{\zeta}_j .$$

We obtain estimates for the claims quota

$$\ddot{\gamma}_j = \frac{\sum_{\ell=0}^{j} \hat{\zeta}_\ell}{\hat{\kappa}} .$$

4.2.4 *The Cape Cod Method*

The Cape Cod method uses the same model as the additive method. However, instead of relying on the risk volume only also the present accumulated payments are included for the prediction of the future payments. One compares the prior predictions of the accumulated payments with the actual payments. That is, one has prior predictions for the payments up till now, which can be compared to the actual payments up till now. The claim quota with respect to the prior predictions becomes

$$\hat{\kappa} = \frac{\sum_{j=0}^{n} C_{j,n-j}}{\sum_{j=0}^{n} \gamma_{n-j} v_j} .$$

Here, we let $C_{i,j} = C_{i,J}$ and $\gamma_j = 1$ for $j > J$. With $\hat{\kappa}$, new predictions for the payments $C_{i,j}$ of year i until year $i + j > n$ can be obtained

$$\hat{C}_{i,j} = C_{i,n-i} + \hat{\kappa}(\gamma_j - \gamma_{n-i}) v_i .$$

The reserves needed are therefore

$$R_i = \hat{\kappa}(\gamma_J - \gamma_{n-1}) v_i .$$

Compared to the chain-ladder method, the Cape Cod method smooths the predictions with respect to outliers. An extraordinary large payment until time $n - i$ will lead to a higher $\hat{\kappa}$. However, the prediction for the reserves is only influenced by the outlier through $\hat{\kappa}$ and not directly, as is the case in the chain-ladder method. In contrast, higher reserves will also be built for the other occurrence years. The method adjusts the prior estimates of the aggregate claims by the factor $\hat{\kappa}$.

4.2.5 The Bornhuetter–Ferguson Method

The Bornhuetter–Ferguson method works in the same way as the loss development method except the risk volume is replaced by an external estimator α_i. The prediction for the accumulated payments $C_{i,j}$ until time $i + j > n$ for occurrence year i is

$$\hat{C}_{i,j} = C_{i,n-i} + (\gamma_j - \gamma_{n-i})\alpha_i .$$

The predictions for the reserves becomes

$$R_i = (\gamma_J - \gamma_{n-i})\alpha_i .$$

As a variant of the method, one could replace α_i with the Bornhuetter–Ferguson predictions $\hat{C}_{i,J}$. With $\hat{C}_{i,J}^{(0)} = \alpha_i$, this yields predictions

$$\hat{C}_{i,j}^{(m+1)} = C_{i,n-i} + (\gamma_j - \gamma_{n-i})\hat{C}_{i,J}^{(m)} .$$

$\hat{C}_{i,j}^{(1)}$ is then the Bornhuetter–Ferguson prediction above. We get in the limit $\lim_{m\to\infty} \hat{C}_{i,j}^{(m)}$ the predictions of the loss-development method.

4.2.6 The Cross-Classified Model

Suppose that the payments $X_{i,j}$ are of the form

$$X_{i,j} = \alpha_i \vartheta_j + \text{error}$$

for some parameters α_i, $i = 1, 2, \ldots, n$, ϑ_j, $j = 0, 1, \ldots, J$. We assume $J \le n-1$ and set $\vartheta_j = 0$ for $j > J$. In order for there to be a unique representation, we assume $\sum_{j=0}^{J} \vartheta_j = 1$. For estimating the paramcters, we use the equations

$$\sum_{j=0}^{n-i} \hat{\alpha}_i \hat{\vartheta}_j = \sum_{j=0}^{n-i} X_{i,j} = C_{i,n-i} \;,$$

$$\sum_{i=1}^{n-j} \hat{\alpha}_i \hat{\vartheta}_j = \sum_{i=1}^{n-j} X_{i,j} \;,$$

$$\sum_{j=0}^{J} \hat{\vartheta}_j = 1 \;.$$

Note that $\hat{\vartheta}_j = 0$ for $j > J$ follows from the equations. It turns out that the method is equivalent to the chain-ladder method.

Lemma 4.1 *Suppose $J \le n-1$. Then $\hat{\alpha}_i$ is the chain-ladder prediction $\hat{C}_{i,J}$ and $\hat{\vartheta}_j$ corresponds to the corresponding prediction in the chain ladder method, $\sum_{\ell=0}^{j} \hat{\vartheta}_\ell / \sum_{\ell=0}^{j-1} \hat{\vartheta}_\ell = \hat{\varphi}_j$.*

Proof We first prove by induction that

$$\sum_{i=1}^{n-j} C_{i,j} = \sum_{i=1}^{n-j} \hat{\alpha}_i \sum_{\ell=0}^{j} \hat{\vartheta}_\ell \;.$$

If $j = n-1$, then we have to prove

$$C_{1,n-1} = \hat{\alpha}_1 \sum_{\ell=0}^{n-1} \hat{\vartheta}_\ell = \hat{\alpha}_1 \;,$$

which is one of the equations to be satisfied. Suppose now that the equation holds for $j = k$. Then

$$\begin{aligned}\sum_{i=1}^{n-(k-1)} C_{i,k-1} &= \sum_{i=1}^{n-k} (C_{i,k} - X_{i,k}) + C_{n-(k-1),k-1} \\ &= \sum_{i=1}^{n-k} \hat{\alpha}_i \sum_{\ell=0}^{k} \hat{\vartheta}_\ell - \sum_{i=1}^{n-k} X_{i,k} + \sum_{\ell=0}^{k-1} \hat{\alpha}_{n-(k-1)} \hat{\vartheta}_\ell\end{aligned}$$

$$= \sum_{i=1}^{n-k} \hat{\alpha}_i \sum_{\ell=0}^{k} \hat{\vartheta}_\ell - \sum_{i=1}^{n-k} \hat{\alpha}_i \hat{\vartheta}_k + \hat{\alpha}_{n-(k-1)} \sum_{\ell=0}^{k-1} \hat{\vartheta}_\ell$$

$$= \sum_{i=1}^{n-k} \hat{\alpha}_i \sum_{\ell=0}^{k-1} \hat{\vartheta}_\ell + \hat{\alpha}_{n-(k-1)} \sum_{\ell=0}^{k-1} \hat{\vartheta}_\ell$$

$$= \sum_{i=1}^{n-(k-1)} \hat{\alpha}_i \sum_{\ell=0}^{k-1} \hat{\vartheta}_\ell \,.$$

Thus the claim also holds for $j = k - 1$. We find further

$$\sum_{i=1}^{n-j} C_{i,j-1} = \sum_{i=1}^{n-j} (C_{i,j} - X_{i,j}) = \sum_{i=1}^{n-j} \hat{\alpha}_i \sum_{\ell=0}^{j} \hat{\vartheta}_\ell - \sum_{i=1}^{n-j} \hat{\alpha}_i \hat{\vartheta}_j = \sum_{i=1}^{n-j} \hat{\alpha}_i \sum_{\ell=0}^{j-1} \hat{\vartheta}_\ell \,.$$

This gives

$$\frac{\sum_{i=1}^{n-j} C_{i,j}}{\sum_{i=1}^{n-j} C_{i,j-1}} = \frac{\sum_{i=1}^{n-j} \hat{\alpha}_i \sum_{\ell=0}^{j} \hat{\vartheta}_\ell}{\sum_{i=1}^{n-j} \hat{\alpha}_i \sum_{\ell=0}^{j-1} \hat{\vartheta}_\ell} = \frac{\sum_{\ell=0}^{j} \hat{\vartheta}_\ell}{\sum_{\ell=0}^{j-1} \hat{\vartheta}_\ell} \,.$$

This means that $\hat{\vartheta}_j$ are the estimators corresponding to the chain-ladder predictions. This proves the result. □

We next consider a model that corresponds to the chain-ladder method or the cross-classified model.

Suppose the payments are independent and Poisson distributed with $\mathbb{E}[X_{i,j}] = \alpha_i \vartheta_j$. We need to estimate the parameters. We use the maximum likelihood method. The log-likelihood is

$$\sum_{i=1}^{n} \sum_{j=0}^{n-i} [X_{i,j} \log(\alpha_i \vartheta_j) - \log(X_{i,j}!) - \alpha_i \vartheta_j] \,.$$

The estimators thus have to solve

$$\sum_{j=0}^{n-i} [X_{i,j}/\hat{\alpha}_i - \hat{\vartheta}_j] = 0 \,,$$

$$\sum_{i=1}^{n-j} [X_{i,j}/\hat{\vartheta}_j - \hat{\alpha}_i] = 0 \,.$$

This implies that the estimators are determined as for the cross-classified model.

Remark 4.2 The cross-classified model is sometimes also called the *marginal sums method*. ■

4.3 The Dirichlet Model

Denote the aggregate claim from occurrence year i by $C_i = C_{i,J}$. The claim size distribution is $\mathbb{P}[C_i \le x] = F_{v_i}(x)$, depending on a volume mass v_i. The amount paid in development year j is $X_{i,j} = \xi_{i,j} C_i$, where $\sum_{j=0}^J \xi_{i,j} = 1$. The $\{(\xi_{i,0}, \xi_{i,1}, \ldots, \xi_{i,J}) : i \ge 1\}$ are assumed to be iid. random vectors. We construct these vectors in the following way.

Let $\gamma_j > 0$ be positive numbers. The random variables $\{Q_{i,j}\}$ are independent and $\Gamma(\gamma_j, 1)$ distributed. The second parameter does not matter because it is a scaling parameter that will finally disappear. By the properties of the Gamma distribution, we have that $\sum_{j=0}^k Q_{i,j}$ is $\Gamma(\sum_{j=0}^k \gamma_j, 1)$ distributed. Let $Q_{i,\cdot} = \sum_{j=0}^J Q_{i,j}$. We let $\xi_{i,j} = Q_{i,j}/Q_{i,\cdot}$ and $\gamma = \sum_{j=0}^J \gamma_j$. Then $\xi_{i,j} \in (0,1)$ and $\sum_{j=0}^J \xi_{i,j} = 1$.

We next show that $\xi_{i,0}, \xi_{i,1}, \ldots, \xi_{i,J}$ is *Dirichlet distributed* with parameters $\gamma_0, \ldots, \gamma_J$.

Lemma 4.3 *The variables $\xi_{i,0}, \xi_{i,1}, \ldots, \xi_{i,J-1}$ have an absolutely continuous joint distribution with density*

$$\Gamma(\gamma) \prod_{j=0}^J \frac{x_j^{\gamma_j - 1}}{\Gamma(\gamma_j)} ,$$

where $x_J = 1 - \sum_{j=0}^{J-1} x_j$, provided $x_j > 0$ for all $0 \le j \le J$.

Remark 4.4 Note that $\xi_{i,J} = 1 - \sum_{j=0}^{J-1} \xi_{i,j}$ by the construction of the variables.

■

Proof The joint density of $\{Q_{i,j} : 0 \le j \le J\}$ is

$$\prod_{j=0}^J \frac{q_j^{\gamma_j - 1}}{\Gamma(\gamma_j)} e^{-q_j} .$$

If we condition on $\{Q_{i,\cdot} = q\}$, we get with $\sum_{j=0}^J q_j = q$ that the joint conditional density of $\{Q_{i,j} : 0 \le k \le J-1\}$ is

$$\frac{\prod_{j=0}^J \frac{q_j^{\gamma_j - 1}}{\Gamma(\gamma_j)} e^{-q_j}}{\frac{q^{\gamma - 1}}{\Gamma(\gamma)} e^{-q}} = \Gamma(\gamma) q^{-J} \prod_{j=0}^J \frac{(q_j/q)^{\gamma_j - 1}}{\Gamma(\gamma_j)} .$$

This yields the conditional joint distribution of the increments of $\{\xi_{i,j} : 0 \le k \le J-1\}$ given $\{Q_{i,\cdot} = q\}$

$$\Gamma(\gamma)\prod_{j=0}^{J}\frac{x_j^{\gamma_j-1}}{\Gamma(\gamma_j)}\;.$$

We see that the density is independent of q. This implies that the unconditional density also has this form. □

We next have to estimate the parameters. We suppose that J is known and that $F_{v_i}(x)$ is absolutely continuous with density $f_{v_i}(x)$, with known densities and risk volumes v_i.

Lemma 4.5 *The variable* $\sum_{j=0}^{k}\xi_{i,j}$ *is* $B(\sum_{j=0}^{k}\gamma_j, \sum_{j=k+1}^{J}\gamma_j)$ *distributed, where* $1 \le k < J$. *The variables* $\{\xi_{i,j}/\sum_{j=0}^{k}\xi_{i,j}\} : 0 \le j \le k$ *are again Dirichlet distributed with parameters* $\gamma_0, \ldots, \gamma_k$ *independent of* $\sum_{j=0}^{k}\xi_{i,j}$.

Proof We prove the first assertion by induction. If $J = 1$, then the Dirichlet distribution is the beta distribution and the assertion is trivial. Suppose the assertion holds for J. We prove it now for $J + 1$. If $k < J$, we can integrate with respect to $\xi_{i,J}$ to get the joint density of $\xi_{i,0}, \ldots, \xi_{i,J-1}$

$$\begin{aligned}\Gamma(\gamma)\prod_{j=0}^{J-1}\frac{x_j^{\gamma_j-1}}{\Gamma(\gamma_j)}&\int_0^{1-\sum_{j=0}^{J-1}x_j}\frac{x_J^{\gamma_J}(1-\sum_{j=0}^{J-1}x_j-x_J)^{\gamma_{J+1}-1}}{\Gamma(\gamma_J)\Gamma(\gamma_{J+1})}\,\mathrm{d}x_J\\ &=\Gamma(\gamma)\frac{(1-\sum_{j=0}^{J-1}x_j)^{\gamma_J+\gamma_{J-1}-1}}{\Gamma(\gamma_J+\gamma_{J-1})}\prod_{j=0}^{J-1}\frac{x_j^{\gamma_j-1}}{\Gamma(\gamma_j)}\;.\end{aligned}$$

Hence, $\xi_{i,0}, \ldots, \xi_{i,J-1}, \xi_{i,J} + \xi_{i,J+1}$ is Dirichlet distributed with parameters $\gamma_0, \ldots, \gamma_{J-1}, \gamma_J+\gamma_{J+1}$. From this, the beta distribution follows from the assumption for J. The analogous argument follows if $k = J$, if we integrate with respect to $\xi_{i,0}$.

For the second assertion we can without loss of generality assume $k = J - 1$. Let $\tilde{\gamma} = \sum_{j=0}^{k}\gamma_j$. The joint conditional density of $\xi_{i,0}, \ldots, \xi_{i,k-1}$ given $\sum_{j=0}^{k}\xi_{i,j} = x$ becomes

$$\frac{\Gamma(\gamma)\prod_{j=0}^{J}\frac{x_j^{\gamma_j-1}}{\Gamma(\gamma_j)}}{\frac{\Gamma(\gamma)}{\Gamma(\tilde{\gamma})\Gamma(\gamma_J)}x^{\gamma-1}x_J^{\gamma_J-1}}=\Gamma(\tilde{\gamma})x^{-k}\prod_{j=0}^{k}\frac{(x_j/x)^{\gamma_j-1}}{\Gamma(\gamma_j)}\;.$$

This yields the desired distribution for $\xi_{i,0}/x, \ldots, \xi_{i,k}/k$

$$\Gamma(\tilde{\gamma})\prod_{j=0}^{k}\frac{x_j^{\gamma_j-1}}{\Gamma(\gamma_j)}\;.$$

We see that the density is independent of x. This proves the assertion. □

The joint density of $C_i, X_{i,0}, \ldots, X_{i,n-i}$ becomes

$$f_{\upsilon_i}(s)\Gamma\left(\sum_{j=0}^{n-i}\gamma_j\right)\frac{(1-\sum_{j=0}^{n-i}z_j/s)^{\sum_{j=n-i+1}^{J}\gamma_j-1}}{\Gamma(\sum_{j=n-i+1}^{J}\gamma_j)s^{n-i+1}}\prod_{j=0}^{n-i}\frac{(z_j/s)^{\gamma_j-1}}{\Gamma(\gamma_j)}\,, \tag{4.1}$$

provided $n-i<J$ and $s>\sum_{j=0}^{n-i}z_j$. Thus, $X_{i,0}, \ldots, X_{i,n-i}$ has the joint density

$$\int_{s_0}^{\infty} f_{\upsilon_i}(s)\Gamma\left(\sum_{j=0}^{n-i}\gamma_j\right)\frac{(1-s_0/s)^{\sum_{j=n-i+1}^{J}\gamma_j-1}}{\Gamma(\sum_{j=n-i+1}^{J}\gamma_j)s^{n-i+1}}\prod_{j=0}^{n-i}\frac{(z_j/s)^{\gamma_j-1}}{\Gamma(\gamma_j)}\,\mathrm{d}s\,,$$

where $s_0=\sum_{j=0}^{n-i}z_j$. If $J\le n-i$, we get the density of $C_i, X_{i,0}, \ldots, X_{i,J-1}$

$$f_{\upsilon_i}(s)\Gamma(\gamma)\frac{1}{s^{n-i}}\prod_{j=0}^{J}\frac{(z_j/s)^{\gamma_j-1}}{\Gamma(\gamma_j)}\,,$$

with $z_J=s-\sum_{j=0}^{J-1}z_j$. Now, the joint density of the observable variables can be written, and the parameters can be determined using the maximum likelihood method. This has to be done numerically.

We next want to predict C_i if $n<J$. We assume that the parameters γ_k are known (or already estimated). The joint density of the observations is given in (4.1). From this, we can find the conditional density of C_i given $X_{i,0}, \ldots, X_{i,n-i}$. Taking the expected value yields the estimate

$$\frac{\displaystyle\int_{s_0}^{\infty} f_{\upsilon_i}(s)\frac{(1-s_0/s)^{\sum_{j=n-i+1}^{J}\gamma_j-1}}{s^{n-i}}\prod_{j=0}^{n-i}(z_j/s)^{\gamma_j-1}\,\mathrm{d}s}{\displaystyle\int_{s_0}^{\infty} f_{\upsilon_i}(s)\frac{(1-s_0/s)^{\sum_{j=n-i+1}^{J}\gamma_j-1}}{s^{n-i+1}}\prod_{j=0}^{n-i}(z_j/s)^{\gamma_j-1}\,\mathrm{d}s}\,.$$

Example 4.6 Suppose that $f_{\upsilon_i}(s)=\alpha(\beta\upsilon_i)^\alpha s^{-\alpha-1}\mathbb{1}_{s>\beta\upsilon_i}$ follows a shifted Pareto distribution. Then for $n-1<J$ the density of $X_{i,0}, \ldots, X_{i,n-i}$ becomes

$$\begin{aligned}&\Gamma\left(\sum_{j=0}^{n-i}\gamma_j\right)(\beta\upsilon_i)^\alpha\int_{s_0\vee\beta\upsilon_i}^{\infty}\frac{(1-s_0/s)^{\sum_{j=n-i+1}^{J}\gamma_j-1}}{\Gamma(\sum_{j=n-i+1}^{J}\gamma_j)s^{n-i+\alpha+2}}\prod_{j=0}^{n-i}\frac{(z_j/s)^{\gamma_j-1}}{\Gamma(\gamma_j)}\,\mathrm{d}s\\ &\quad=\frac{\Gamma\left(\sum_{j=0}^{n-i}\gamma_j\right)}{\Gamma(\sum_{j=n-i+1}^{J}\gamma_j)}\prod_{j=0}^{n-i}\frac{(z_j)^{\gamma_j-1}}{\Gamma(\gamma_j)}(\beta\upsilon_i)^\alpha\end{aligned}$$

$$\times \int_{s_0 \vee \beta v_i}^{\infty} (s - s_0)^{\sum_{j=n-i+1}^{J} \gamma_j - 1} s^{-\gamma-\alpha-1} \, \mathrm{d}s$$

$$= s_0^{-\alpha - \sum_{j=0}^{n-i} \gamma_j} \frac{\Gamma\left(\sum_{j=0}^{n-i} \gamma_j\right)}{\Gamma(\sum_{j=n-i+1}^{J} \gamma_j)} \prod_{j=0}^{n-i} \frac{(z_j)^{\gamma_j - 1}}{\Gamma(\gamma_j)} (\beta v_i)^{\alpha}$$

$$\times \int_{1 \vee \beta v_i / s_0}^{\infty} (y - 1)^{\sum_{j=n-i+1}^{J} \gamma_j - 1} y^{-\gamma-\alpha-1} \, \mathrm{d}y \, .$$

For the estimate of C_i, we find

$$\frac{\int_{1 \vee \beta v_i / s_0}^{\infty} (y - 1)^{\sum_{j=n-i+1}^{J} \gamma_j - 1} y^{-\gamma-\alpha} \, \mathrm{d}y}{\int_{1 \vee \beta v_i / s_0}^{\infty} (y - 1)^{\sum_{j=n-i+1}^{J} \gamma_j - 1} y^{-\gamma-\alpha-1} \, \mathrm{d}y} s_0 \, .$$

A simple explicit expression is found if $s_0 \geq \beta v_i$,

$$\frac{\Gamma(\sum_{j=n-i+1}^{J} \gamma_j) \Gamma(\alpha + \sum_{j=0}^{n-i} \gamma_j) \Gamma(\gamma + \alpha + 1)}{\Gamma(\gamma + \alpha) \Gamma(\sum_{j=n-i+1}^{J} \gamma_j) \Gamma(\alpha + 1 + \sum_{j=0}^{n-i} \gamma_j)} s_0 = \frac{\gamma + \alpha}{\alpha + \sum_{j=0}^{n-i} \gamma_j} s_0 \, .$$

■

Bibliographical Remarks

Further literature on claims reserving can be found in [28, 70, 88, 100, 108, 127]. In these books, the reader will also find extensions (for example to the multivariate case) and generalisations of the methods. An overview of linear prediction models is found in [38]. A comprehensive survey including the parameter uncertainty aspects is found in [109] or [140], see also the references therein. Solvency aspects are addressed in [141]. The oldest reference to the chain-ladder method the author is aware of is [126]. The estimation error is addressed in [99]. The Bornhuetter–Ferguson method was introduced in [25]. It is also called the *Benktander–Hovinen* method, see also [101, 102]. The cross-classified model originates from [81], see also [93, 98, 128]. The Poisson example of Sect. 4.2.6 goes back to [81]. The Dirichlet model was introduced in [60]. Here we follow [104]. A Bayesian approach is described in [57].

Chapter 5
The Cramér–Lundberg Model

To model the surplus of an insurance portfolio we use a stochastic process which is, besides Brownian motion, the simplest possible. In particular, we will be interested in ruin probabilities. We show that many characteristics can be calculated such as, for example, the capital prior to and immediately after ruin. At the end we discuss an alternative measure for the risk that not only takes the event of ruin into account but also the time to ruin and the deficit to ruin.

5.1 Definition of the Cramér–Lundberg Process

We have seen that the compound Poisson model has nice properties. For instance, it can be derived as a limit of individual models. This was the motivation for Filip Lundberg [96] to postulate a continuous time risk model where the aggregate claims in any interval have a compound Poisson distribution. Moreover, the premium income should be modelled. In a portfolio of insurance contracts the premium payments will be spread all over the year. Thus Lundberg assumed that the premium income is continuous over time and that the premium income in any time interval is proportional to the interval length. This leads to the following model for the surplus of an insurance portfolio

$$C_t = u + ct - \sum_{i=1}^{N_t} Y_i\,.$$

u is the initial capital, c is the premium rate. The number of claims in $(0, t]$ is a Poisson process $\{N_t\}$ with rate λ. The claim sizes $\{Y_i\}$ are a sequence of iid. positive random variables independent of $\{N_t\}$. This model is called the *Cramér–Lundberg process* or *classical risk process*.

H. Schmidli, *Risk Theory*, Springer Actuarial,
DOI:10.1007/978-3-319-72005-0_5

We denote the distribution function of the claims by G, its moments by $\mu_n = \mathbb{E}[Y_1^n]$ and its moment generating function by $M_Y(r) = \mathbb{E}[\exp\{rY_1\}]$. Let $\mu = \mu_1$. We assume that $\mu < \infty$. Otherwise, no insurance company would insure such a risk. Note that $G(x) = 0$ for $x < 0$. We will see later that it is no restriction to assume that $G(0) = 0$.

For an insurance company it is important that $\{C_t\}$ stays above a certain level. This solvency level is given by legal restrictions. By adjusting the initial capital there is no loss of generality in assuming this level to be 0. We define the ruin time

$$\tau = \inf\{t > 0 : C_t < 0\}, \qquad (\inf \emptyset = \infty).$$

We will mostly be interested in the probability of ruin in a time interval $(0, t]$

$$\psi(u, t) = \mathbb{P}[\tau \le t \mid C_0 = u] = \mathbb{P}\Big[\inf_{0<s\le t} C_s < 0 \mid C_0 = u\Big]$$

and the probability of ultimate ruin

$$\psi(u) = \lim_{t\to\infty} \psi(u, t) = \mathbb{P}\Big[\inf_{t>0} C_t < 0 \mid C_0 = u\Big].$$

It is easy to see that $\psi(u, t)$ is decreasing in u and increasing in t.

We denote the claim times by $T_1, T_2, \ldots$ and by convention $T_0 = 0$. Let $X_i = c(T_i - T_{i-1}) - Y_i$. If we only consider the process at the claim times we can see that

$$C_{T_n} = u + \sum_{i=1}^{n} X_i$$

is a random walk. Note that $\psi(u) = \mathbb{P}[\inf_{n\in\mathbb{N}} C_{T_n} < 0]$. From the theory of random walks we can see that ruin occurs a.s. iff $\mathbb{E}[X_i] \le 0$ (compare with Lemma E.1 or [59, p. 396]). Hence we will assume in the sequel that

$$\mathbb{E}[X_i] > 0 \iff c\frac{1}{\lambda} - \mu > 0 \iff c > \lambda\mu \iff \mathbb{E}[C_t - u] > 0.$$

Recall that

$$\mathbb{E}\Big[\sum_{i=1}^{N_t} Y_i\Big] = \lambda t \mu.$$

The condition can be interpreted as the average income being strictly larger than the average outflow. Therefore the condition is also called the *net profit condition*.

If the net profit condition is fulfilled then C_{T_n} tends to infinity as $n \to \infty$. Hence

$$\inf\{C_t - u : t > 0\} = \inf\{C_{T_n} - u : n \geq 1\}$$

is a.s. finite. So we can conclude that

$$\lim_{u\to\infty} \psi(u) = 0\,.$$

5.2 A Note on the Model and Reality

In reality the average number of claims in an interval will not be the same all the time. There will be a claim rate $\lambda(t)$ which may be periodic in time. Moreover, the number of individual contracts in the portfolio may vary with time. Let $a(t)$ be the volume of the portfolio at time t. Then the claim number process $\{N_t\}$ is an inhomogeneous Poisson process with rate $a(t)\lambda(t)$. Let

$$\Lambda(t) = \int_0^t a(s)\lambda(s)\,\mathrm{d}s$$

and $\Lambda^{-1}(t)$ be its inverse function. Then $\tilde{N}_t = N_{\Lambda^{-1}(t)}$ is a Poisson process with rate 1. Now let the premium rate vary with t so that $c_t = ca(t)\lambda(t)$ for some constant c. This assumption is natural for changes in the risk volume. It is artificial for the changes in the intensity. For instance, we assume that the company gets more new customers at times with a higher intensity. This effect may arise because customers withdraw from their old insurance contracts and write new contracts with another company after claims occurred because they were not satisfied by the handling of claims by their old companies.

The premium income in the interval $(0, t]$ is $c\Lambda(t)$. Let $\tilde{C}_t = C_{\Lambda^{-1}(t)}$. Then

$$\tilde{C}_t = u + c\Lambda(\Lambda^{-1}(t)) - \sum_{i=1}^{N_{\Lambda^{-1}(t)}} Y_i = u + ct - \sum_{i=1}^{\tilde{N}_t} Y_i$$

is a Cramér–Lundberg process. Thus we should not consider time to be the real time but *operational time*.

The event of ruin almost never occurs in practice. If an insurance company observes that their surplus is decreasing they will immediately increase their premia. On the other hand an insurance company is built up on different portfolios. Ruin in one portfolio does not mean bankruptcy. Therefore ruin is only a technical term. The surplus will also be a technical term in practice. If the business is going well then the share holders will decide to get a higher dividend. To model this we would have to assume a premium rate dependent on the surplus, but then it would be hard to obtain any useful results. For some references see, for instance, [13, 63].

The probability of ruin should be regarded as a measure of the risk and is used for decision taking; for example, the premium calculation or the computation of reinsurance retention levels. One therefore freezes the present portfolio and looks at how the surplus would behave in the future. For an actuary it is important to be able to take a good decision in reasonable time. Therefore it is fine to simplify the model in order to be able to check whether a decision has the desired effect or not.

In Chap. 1 on risk models we saw that a negative binomial distribution for the number of claims in a certain time interval would be preferable. We will later consider such models. However, actuaries often use the Cramér–Lundberg model for their calculations, even though there is a certain lack of realism. The reason is that this model is well understood, and the basic properties of a risk also are present in this simple model.

5.3 A Differential Equation for the Ruin Probability

We first prove that $\{C_t\}$ is a strong Markov process.

Lemma 5.1 *Let $\{C_t\}$ be a Cramér–Lundberg process and T be a finite stopping time. Then the stochastic process $\{C_{T+t} - C_T : t \geq 0\}$ is a Cramér–Lundberg process with initial capital 0 and is independent of $\mathcal{F}_T$.*

Proof We can write $C_{T+t} - C_T$ as

$$ct - \sum_{i=N_T+1}^{N_{T+t}} Y_i \,.$$

Because the claim amounts are iid. and independent of $\{N_t\}$ we only have to prove that $\{N_{T+t} - N_T\}$ is a Poisson process independent of $\mathcal{F}_T$. Because $\{N_t\}$ is a renewal process it is enough to show that $T_{N_T+1} - T$ is Exp(λ) distributed and independent of $\mathcal{F}_T$. Condition on T_{N_T} and T. Then

$$\begin{aligned}
&\mathbb{P}[T_{N_T+1} - T > x \mid T_{N_T}, T] \\
&\quad = \mathbb{P}[T_{N_T+1} - T_{N_T} > x + T - T_{N_T} \mid T_{N_T}, T, T_{N_T+1} - T_{N_T} > T - T_{N_T}] \\
&\quad = \mathrm{e}^{-\lambda x}
\end{aligned}$$

by the lack of memory property of the exponential distribution. The assertion follows because T_{N_T+1} depends on $\mathcal{F}_T$ via T_{N_T} and T only. The latter, even though intuitively clear, follows readily noting that $\mathcal{F}_T$ is generated by sets of the form $\{N_T = n, A_n\}$ where $n \in \mathbb{N}$ and $A_n \in \sigma(Y_1, \ldots, Y_n, T_1, \ldots, T_n)$. □

Let h be small. If ruin does not occur in the interval $(0, T_1 \wedge h]$ then a new Cramér–Lundberg process starts at time $T_1 \wedge h$ with new initial capital $C_{T_1 \wedge h}$. Let $\delta(u) = 1 - \psi(u)$ denote the survival probability. Using that the interarrival times are exponentially distributed we have the density $\lambda e^{-\lambda t}$ of the distribution of T_1 and $\mathbb{P}[T_1 > h] = e^{-\lambda h}$. Noting that $\delta(x) = 0$ for $x < 0$, we obtain

$$\delta(u) = e^{-\lambda h}\delta(u + ch) + \int_0^h \int_0^{u+ct} \delta(u + ct - y)\, dG(y)\lambda e^{-\lambda t}\, dt\ .$$

Letting h tend to 0 shows that $\delta(u)$ is right continuous. Rearranging the terms and dividing by h yields

$$c\,\frac{\delta(u + ch) - \delta(u)}{ch} = \frac{1 - e^{-\lambda h}}{h}\delta(u + ch) - \frac{1}{h}\int_0^h \int_0^{u+ct} \delta(u + ct - y)\, dG(y)\lambda e^{-\lambda t}\, dt\ .$$

Letting h tend to 0 shows that $\delta(u)$ is differentiable from the right and

$$c\delta'(u) = \lambda\Bigl[\delta(u) - \int_0^u \delta(u - y)\, dG(y)\Bigr]. \tag{5.1}$$

Replacing u by $u - ch$ gives

$$\delta(u - ch) = e^{-\lambda h}\delta(u) + \int_0^h \int_0^{u-c(h-t)} \delta(u - (c - h)t - y)\, dG(y)\lambda e^{-\lambda t}\, dt\ .$$

We conclude that $\delta(u)$ is continuous. Rearranging the terms, dividing by h and letting $h \downarrow 0$ yields the equation

$$c\delta'(u) = \lambda\Bigl[\delta(u) - \int_0^{u-} \delta(u - y)\, dG(y)\Bigr].$$

We see that $\delta(u)$ is differentiable at all points u where $G(u)$ is continuous. If $G(u)$ has a jump then the derivatives from the left and from the right do not coincide. We have shown that $\delta(u)$ is absolutely continuous, because $G(u)$ can have at most countably many jumps. If we now let $\delta'(u)$ be the derivative from the right we have that $\delta'(u)$ is a density of $\delta(u)$.

The difficulty with Eq. (5.1) is that it contains both the derivative of $\delta(u)$ and an integral. Let us try to get rid of the derivative. We find

$$
\begin{aligned}
\frac{c}{\lambda}(\delta(u)-\delta(0)) &= \frac{1}{\lambda}\int_0^u c\delta'(x)\,\mathrm{d}x = \int_0^u \delta(x)\,\mathrm{d}x - \int_0^u\int_0^x \delta(x-y)\,\mathrm{d}G(y)\,\mathrm{d}x \\
&= \int_0^u \delta(x)\,\mathrm{d}x - \int_0^u\int_y^u \delta(x-y)\,\mathrm{d}x\,\mathrm{d}G(y) \\
&= \int_0^u \delta(x)\,\mathrm{d}x - \int_0^u\int_0^{u-y} \delta(x)\,\mathrm{d}x\,\mathrm{d}G(y) \\
&= \int_0^u \delta(x)\,\mathrm{d}x - \int_0^u\int_0^{u-x} \mathrm{d}G(y)\,\delta(x)\,\mathrm{d}x \\
&= \int_0^u \delta(x)(1-G(u-x))\,\mathrm{d}x = \int_0^u \delta(u-x)(1-G(x))\,\mathrm{d}x\,.
\end{aligned}
$$

Note that $\int_0^\infty (1-G(x))\,\mathrm{d}x = \mu$. Letting $u \to \infty$ we can, by the bounded convergence theorem, interchange limit and integral and get

$$
c(1-\delta(0)) = \lambda \int_0^\infty (1-G(x))\,\mathrm{d}x = \lambda\mu\,,
$$

where we used that $\delta(u) \to 1$. It follows that

$$
\delta(0) = 1 - \frac{\lambda\mu}{c}, \qquad \psi(0) = \frac{\lambda\mu}{c}\,.
$$

Replacing $\delta(u)$ by $1-\psi(u)$ we obtain

$$
\begin{aligned}
c\psi(u) &= \lambda\mu - \lambda\int_0^u (1-\psi(u-x))(1-G(x))\,\mathrm{d}x \\
&= \lambda\Bigl(\int_u^\infty (1-G(x))\,\mathrm{d}x + \int_0^u \psi(u-x)(1-G(x))\,\mathrm{d}x\Bigr)\,. \qquad (5.2)
\end{aligned}
$$

In Sect. 5.8 we shall obtain a natural interpretation of (5.2).

Example 5.2 Let the claims be $\operatorname{Exp}(\alpha)$ distributed. Then Eq. (5.1) can be written as

$$
c\delta'(u) = \lambda\Bigl[\delta(u) - \mathrm{e}^{-\alpha u}\int_0^u \delta(y)\alpha\mathrm{e}^{\alpha y}\,\mathrm{d}y\Bigr]\,.
$$

Differentiating yields

$$
c\delta''(u) = \lambda\Bigl[\delta'(u) + \alpha\mathrm{e}^{-\alpha u}\int_0^u \delta(y)\alpha\mathrm{e}^{\alpha y}\,\mathrm{d}y - \alpha\delta(u)\Bigr] = \lambda\delta'(u) - \alpha c\delta'(u)\,.
$$

The solution to this differential equation is

$$
\delta(u) = A + B\mathrm{e}^{-(\alpha-\lambda/c)u}\,.
$$

Because $\delta(u) \to 1$ as $u \to \infty$ we get $A = 1$. Because $\delta(0) = 1 - \lambda/(\alpha c)$ the solution is

$$\delta(u) = 1 - \frac{\lambda}{\alpha c} e^{-(\alpha-\lambda/c)u}$$

or

$$\psi(u) = \frac{\lambda}{\alpha c} e^{-(\alpha-\lambda/c)u} .$$

■

5.4 The Adjustment Coefficient

Let

$$\theta(r) = \lambda(M_Y(r) - 1) - cr , \tag{5.3}$$

provided $M_Y(r)$ exists. Then we find the following martingale.

Lemma 5.3 *Let $r \in \mathbb{R}$ such that $M_Y(r) < \infty$. Then the stochastic process*

$$\{\exp\{-rC_t - \theta(r)t\}\} \tag{5.4}$$

is a martingale.

Proof By the Markov property we have for $s < t$

$$\begin{aligned}\mathbb{E}[e^{-rC_t-\theta(r)t} \mid \mathcal{F}_s] &= \mathbb{E}[e^{-r(C_t-C_s)}]e^{-rC_s-(\lambda(M_Y(r)-1)-cr)t} \\ &= \mathbb{E}[e^{r\sum_{N_s+1}^{N_t} Y_i}]e^{-rC_s-\lambda(M_Y(r)-1)t+crs} \\ &= e^{-rC_s-\lambda(M_Y(r)-1)s+crs} = e^{-rC_s-\theta(r)s} .\end{aligned}$$

□

It would be nice to have a martingale that only depends on the surplus and not explicitly on time. Let us therefore consider the equation $\theta(r) = 0$. This equation obviously has the solution $r = 0$. We differentiate $\theta(r)$.

$$\theta'(r) = \lambda M_Y'(r) - c ,$$

$$\theta''(r) = \lambda M_Y''(r) = \lambda \mathbb{E}[Y^2 e^{rY}] > 0 .$$

The function $\theta(r)$ is strictly convex. For $r = 0$

$$\theta'(0) = \lambda M_Y'(0) - c = \lambda\mu - c < 0$$

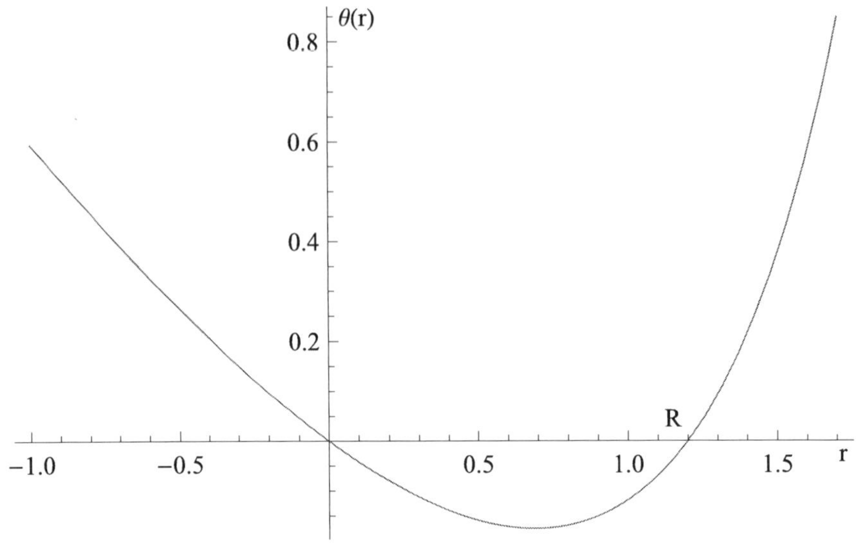

Fig. 5.1 The function $\theta(r)$ and the adjustment coefficient R

by the net profit condition. There might be at most one additional solution R to the equation $\theta(R) = 0$ and $R > 0$, see Fig. 5.1. If this solution exists we call it the *adjustment coefficient* or the *Lundberg exponent*. The Lundberg exponent will play an important rôle in the estimation of the ruin probabilities.

Example 5.4 For Exp(α) distributed claims we have to solve

$$\lambda\Big(\frac{\alpha}{\alpha - r} - 1\Big) - cr = \frac{\lambda r}{\alpha - r} - cr = 0\,.$$

For $r \neq 0$ we find $R = \alpha - \lambda/c$. Note that

$$\psi(u) = \frac{\lambda}{\alpha c}\mathrm{e}^{-Ru}\,.$$

∎

In general the adjustment coefficient is hard to calculate. So we try to find some bounds for the adjustment coefficient. Note that the second moment μ_2 exists if the Lundberg exponent exists. Consider the function $\theta(r)$ for $r > 0$.

$$\begin{aligned}
\theta''(r) &= \lambda\mathbb{E}[Y^2\mathrm{e}^{rY}] > \lambda\mathbb{E}[Y^2] = \lambda\mu_2\,,\\
\theta'(r) &= \theta'(0) + \int_0^r \theta''(s)\,\mathrm{d}s > -(c - \lambda\mu) + \lambda\mu_2 r\,,\\
\theta(r) &= \theta(0) + \int_0^r \theta'(s)\,\mathrm{d}s > \lambda\mu_2\frac{r^2}{2} - (c - \lambda\mu)r\,.
\end{aligned}$$

The last inequality yields for $r = R$

$$0 = \theta(R) > R\Bigl(\lambda\mu_2 \frac{R}{2} - (c - \lambda\mu)\Bigr)$$

from which an upper bound of R

$$R < \frac{2(c - \lambda\mu)}{\lambda\mu_2}$$

follows.

We are not able to find a lower bound in general, but in the case of bounded claims we get a lower bound for R. Assume that $Y_1 \le M$ a.s.

Let us first consider the function

$$f(x) = \frac{x}{M}\left(e^{RM} - 1\right) - \left(e^{Rx} - 1\right).$$

Its second derivative is

$$f''(x) = -R^2 e^{Rx} < 0 .$$

$f(x)$ is concave with $f(0) = f(M) = 0$. Thus $f(x) > 0$ for $0 < x < M$. The function $h(x) = xe^x - e^x + 1$ has a minimum at 0. Thus $h(x) > h(0) = 0$ for $x \ne 0$. In particular

$$\frac{1}{RM}\left(e^{RM} - 1\right) < e^{RM} .$$

We calculate

$$\begin{aligned} M_Y(R) - 1 &= \int_0^M \left(e^{Rx} - 1\right) \, dG(x) < \int_0^M \frac{x}{M}\left(e^{RM} - 1\right) \, dG(x) \\ &= \frac{\mu}{M}\left(e^{RM} - 1\right). \end{aligned}$$

From the equation determining R we get

$$0 = \lambda(M_Y(R) - 1) - cR < \frac{\lambda\mu}{M}\left(e^{RM} - 1\right) - cR < \lambda\mu R e^{RM} - cR .$$

It follows readily that

$$R > \frac{1}{M} \log \frac{c}{\lambda\mu} .$$

5.5 Lundberg's Inequality

We will now connect the adjustment coefficient and the ruin probabilities.

Theorem 5.5 *Assume that the adjustment coefficient R exists. Then*

$$\psi(u) < \mathrm{e}^{-Ru} .$$

Proof Assume that the theorem does not hold. Let

$$u_0 = \inf\{u \ge 0 : \psi(u) \ge \mathrm{e}^{-Ru}\} .$$

Because $\psi(u)$ is continuous we get

$$\psi(u_0) = \mathrm{e}^{-Ru_0} .$$

Because $\psi(0) < 1$ we conclude that $u_0 > 0$. Consider Eq. (5.2) for $u = u_0$. We have

$$\begin{aligned} c\psi(u_0) &= \lambda\Bigl[\int_{u_0}^\infty (1 - G(x))\,\mathrm{d}x + \int_0^{u_0} \psi(u_0 - x)(1 - G(x))\,\mathrm{d}x\Bigr] \\ &< \lambda\Bigl[\int_{u_0}^\infty (1 - G(x))\,\mathrm{d}x + \int_0^{u_0} \mathrm{e}^{-R(u_0-x)}(1 - G(x))\,\mathrm{d}x\Bigr] \\ &\le \lambda \int_0^\infty \mathrm{e}^{-R(u_0-x)}(1 - G(x))\,\mathrm{d}x = \lambda \mathrm{e}^{-Ru_0} \int_0^\infty \int_x^\infty \mathrm{e}^{Rx}\,\mathrm{d}G(y)\,\mathrm{d}x \\ &= \lambda \mathrm{e}^{-Ru_0} \int_0^\infty \int_0^y \mathrm{e}^{Rx}\,\mathrm{d}x\,\mathrm{d}G(y) = \lambda \mathrm{e}^{-Ru_0} \int_0^\infty \frac{1}{R}(\mathrm{e}^{Ry} - 1)\,\mathrm{d}G(y) \\ &= \frac{\lambda}{R}\mathrm{e}^{-Ru_0}(M_Y(R) - 1) = c\mathrm{e}^{-Ru_0} , \end{aligned}$$

which is a contradiction. This proves the theorem. □

We now give an alternative proof of the theorem. We will use the positive martingale (5.4) for $r = R$. By the stopping theorem

$$\mathrm{e}^{-Ru} = \mathrm{e}^{-RC_0} = \mathbb{E}[\mathrm{e}^{-RC_{\tau\wedge t}}] \ge \mathbb{E}[\mathrm{e}^{-RC_{\tau\wedge t}}; \tau \le t] = \mathbb{E}[\mathrm{e}^{-RC_\tau}; \tau \le t] .$$

Letting t tend to infinity yields by monotone convergence

$$\mathrm{e}^{-Ru} \ge \mathbb{E}[\mathrm{e}^{-RC_\tau}; \tau < \infty] > \mathbb{P}[\tau < \infty] = \psi(u) \tag{5.5}$$

because $C_\tau < 0$. We have got an upper bound for the ruin probability. We would like to know whether R is the best possible exponent in an exponential upper bound. This question will be answered in the next section.

Example 5.6 Let $G(x) = 1 - pe^{-\alpha x} - (1-p)e^{-\beta x}$, where $0 < \alpha < \beta$ and $0 < p \le \beta(\beta-\alpha)^{-1}$. Note that the expected value is $\frac{p}{\alpha} + \frac{1-p}{\beta}$ and thus

$$c > \frac{\lambda p}{\alpha} + \frac{\lambda(1-p)}{\beta}\,.$$

For $r < \alpha$

$$M_Y(r) = \int_0^\infty e^{rx}\Bigl(\alpha p e^{-\alpha x} + \beta(1-p)e^{-\beta x}\Bigr)\,dx = \frac{\alpha p}{\alpha - r} + \frac{\beta(1-p)}{\beta - r}\,.$$

Thus, we have to solve

$$\lambda\Bigl(\frac{\alpha p}{\alpha - r} + \frac{\beta(1-p)}{\beta - r} - 1\Bigr) - cr = \lambda\Bigl(\frac{pr}{\alpha - r} + \frac{(1-p)r}{\beta - r}\Bigr) - cr = 0\,.$$

We find the obvious solution $r = 0$. If $r \neq 0$ then

$$\lambda p(\beta - r) + \lambda(1-p)(\alpha - r) = c(\alpha - r)(\beta - r)$$

or equivalently

$$cr^2 - ((\alpha+\beta)c - \lambda)r + \alpha\beta c - \lambda((1-p)\alpha + p\beta) = 0\,. \tag{5.6}$$

The solution is

$$\begin{aligned} r &= \frac{1}{2}\Bigl(\alpha + \beta - \frac{\lambda}{c} \pm \sqrt{\Bigl(\alpha+\beta-\frac{\lambda}{c}\Bigr)^2 - 4\Bigl(\alpha\beta - \frac{\lambda}{c}p\beta - \frac{\lambda}{c}(1-p)\alpha\Bigr)}\Bigr) \\ &= \frac{1}{2}\Bigl(\alpha + \beta - \frac{\lambda}{c} \pm \sqrt{\Bigl(\beta-\alpha-\frac{\lambda}{c}\Bigr)^2 + 4p\frac{\lambda}{c}(\beta-\alpha)}\Bigr)\,. \end{aligned}$$

Now we have three solutions, but there should only be two. Note that

$$\alpha\beta - \frac{\lambda}{c}p\beta - \frac{\lambda}{c}(1-p)\alpha = \frac{\alpha\beta}{c}\Bigl(c - \lambda\Bigl(\frac{p}{\alpha} + \frac{1-p}{\beta}\Bigr)\Bigr) > 0$$

and thus both solutions are positive. The larger of the solutions can be written as

$$\alpha + \frac{1}{2}\Bigl(\beta - \alpha - \frac{\lambda}{c} + \sqrt{\Bigl(\beta-\alpha-\frac{\lambda}{c}\Bigr)^2 + 4p\frac{\lambda}{c}(\beta-\alpha)}\Bigr)$$

and is thus larger than α. However, the moment generating function does not exist for $r \ge \alpha$. Thus

$$R = \frac{1}{2}\Bigl(\alpha + \beta - \frac{\lambda}{c} - \sqrt{\Bigl(\beta - \alpha - \frac{\lambda}{c}\Bigr)^2 + 4p\frac{\lambda}{c}(\beta - \alpha)}\Bigr).$$

From Lundberg's inequality it follows that

$$\psi(u) < \mathrm{e}^{-Ru}.$$

∎

5.6 The Cramér–Lundberg Approximation

Consider Eq. (5.2). This equation looks almost like a renewal equation, but

$$\int_0^\infty \frac{\lambda}{c}(1 - G(x))\,\mathrm{d}x = \frac{\lambda\mu}{c} < 1$$

is not a probability distribution. We suppose that the adjustment coefficient exists. Then (5.2) can be written as

$$\psi(u)\mathrm{e}^{Ru} = \int_0^u \psi(u - x)\mathrm{e}^{R(u-x)}\frac{\lambda}{c}(1 - G(x))\mathrm{e}^{Rx}\,\mathrm{d}x + \mathrm{e}^{Ru}\int_u^\infty \frac{\lambda}{c}(1 - G(x))\,\mathrm{d}x.$$

Since

$$\begin{aligned}\int_0^\infty \frac{\lambda}{c}(1 - G(x))\mathrm{e}^{Rx}\,\mathrm{d}x &= \frac{\lambda}{c}\int_0^\infty\int_x^\infty \mathrm{e}^{Rx}\,\mathrm{d}G(y)\,\mathrm{d}x \\ &= \frac{\lambda}{c}\int_0^\infty\int_0^y \mathrm{e}^{Rx}\,\mathrm{d}x\,\mathrm{d}G(y) = \frac{\lambda(M_Y(R) - 1)}{cR} \\ &= 1,\end{aligned} \tag{5.7}$$

we obtain a proper renewal equation. For the application of the key renewal theorem Proposition C.16, we further need that

$$\int_0^\infty x\mathrm{e}^{Rx}\frac{\lambda}{c}(1 - G(x))\,\mathrm{d}x < \infty, \tag{5.8}$$

which is equivalent to $M_Y'(R) < \infty$, see below. It follows from Lemma C.14 that

$$\mathrm{e}^{Ru}\int_u^\infty \frac{\lambda}{c}(1 - G(x))\,\mathrm{d}x$$

is directly Riemann integrable.

Theorem 5.7 *Assume that the Lundberg exponent exists and that* (5.8) *is fulfilled. Then*

$$\lim_{u\to\infty} \psi(u)\mathrm{e}^{Ru} = \frac{c-\lambda\mu}{\lambda M_Y'(R)-c}\,.$$

Proof It only remains to compute the limit.

$$\begin{aligned}\int_0^\infty \mathrm{e}^{Ru}\int_u^\infty \frac{\lambda}{c}(1-G(x))\,\mathrm{d}x\,\mathrm{d}u &= \frac{\lambda}{c}\int_0^\infty\int_0^x \mathrm{e}^{Ru}\,\mathrm{d}u\,(1-G(x))\,\mathrm{d}x\\ &= \frac{\lambda}{cR}\int_0^\infty (\mathrm{e}^{Rx}-1)(1-G(x))\,\mathrm{d}x\\ &= \frac{1}{R}-\frac{\lambda\mu}{cR} = \frac{1}{cR}(c-\lambda\mu)\,,\end{aligned}$$

where we used (5.7). The expected value of the distribution is

$$\begin{aligned}\int_0^\infty x\mathrm{e}^{Rx}\frac{\lambda}{c}(1-G(x))\,\mathrm{d}x &= \frac{\lambda}{c}\int_0^\infty\int_x^\infty x\mathrm{e}^{Rx}\,\mathrm{d}G(y)\,\mathrm{d}x\\ &= \frac{\lambda}{c}\int_0^\infty\int_0^y x\mathrm{e}^{Rx}\,\mathrm{d}x\,\mathrm{d}G(y) = \frac{\lambda}{cR^2}\int_0^\infty (Ry\mathrm{e}^{Ry}-\mathrm{e}^{Ry}+1)\,\mathrm{d}G(y)\\ &= \frac{\lambda(RM_Y'(R)-M_Y(R)+1)}{cR^2} = \frac{\lambda RM_Y'(R)-cR}{cR^2} = \frac{\lambda M_Y'(R)-c}{cR}\,.\end{aligned}$$

The limit value follows readily. □

Remark 5.8 The theorem shows that it is not possible to obtain an exponential upper bound for the ruin probability with an exponent strictly larger than R. ■

The theorem can be written in the following form

$$\psi(u) \sim \frac{c-\lambda\mu}{\lambda M_Y'(R)-c}\mathrm{e}^{-Ru}\,.$$

Thus for large u we get an approximation to $\psi(u)$. This approximation is called the *Cramér–Lundberg approximation*.

Example 5.9 Let the claims be Exp(α) distributed. From $M_Y(r)$ we get that

$$M_Y'(r) = \frac{\alpha}{(\alpha-r)^2}$$

and thus

$$\lim_{u\to\infty}\psi(u)\mathrm{e}^{Ru} = \frac{c-\frac{\lambda}{\alpha}}{\frac{\lambda\alpha}{(\alpha-R)^2}-c} = \frac{c-\frac{\lambda}{\alpha}}{\frac{\lambda\alpha}{\left(\frac{\lambda}{c}\right)^2}-c} = \frac{\lambda}{\alpha c}\frac{\alpha c-\lambda}{\alpha c-\lambda} = \frac{\lambda}{\alpha c}\,.$$

Hence the Cramér–Lundberg approximation

$$\psi(u) \sim \frac{\lambda}{\alpha c} \mathrm{e}^{-(\alpha-\lambda/c)u}$$

becomes exact in this case. ■

Example 5.10 Let $c = \lambda = 1$ and $G(x) = 1 - \frac{1}{3}(\mathrm{e}^{-x} + \mathrm{e}^{-2x} + \mathrm{e}^{-3x})$. The expected value of claim sizes is $\mu = 0.611111$, i.e. the net profit condition is fulfilled. One can show that

$$\psi(u) = 0.550790\,\mathrm{e}^{-0.485131u} + 0.0436979\,\mathrm{e}^{-1.72235u} + 0.0166231\,\mathrm{e}^{-2.79252u}\,.$$

From Theorem 5.7 it follows that $app(u) = 0.550790\,\mathrm{e}^{-0.485131u}$ is the Cramér–Lundberg approximation to $\psi(u)$. Table 5.1 shows the ruin function $\psi(u)$, its Cramér–Lundberg approximation $app(u)$ and the relative error $(app(u) - \psi(u))/\psi(u)$ multiplied by 100 (Er). Note that the relative error is below 1% for $u \ge 1.71358 = 2.8\mu$. ■

Example 5.11 Suppose that $\psi(u) = A\mathrm{e}^{-Ru}$, that is, the Cramér–Lundberg approximation is exact. Then $A = \lambda\mu/c$. (5.1) then reads

$$\begin{aligned} c\mu R &= c\mathrm{e}^{Ru} - \lambda\mu - cG(u)\mathrm{e}^{Ru} + \lambda\mu \int_0^u \mathrm{e}^{Ry}\,\mathrm{d}G(y) \\ &= c\mathrm{e}^{Ru} - \lambda\mu - (c - \lambda\mu)G(u)\mathrm{e}^{Ru} - \lambda\mu \int_0^u R\mathrm{e}^{Ry} G(y)\,\mathrm{d}y\,. \end{aligned}$$

We conclude that $G(u)$ is continuously differentiable. Denote the density by $g(y)$. Taking the derivative,

$$0 = cR\mathrm{e}^{Ru} - (c - \lambda\mu)(g(u) + RG(u))\mathrm{e}^{Ru} - \lambda\mu RG(u)\mathrm{e}^{Ru}\,.$$

Table 5.1 Cramér–Lundberg approximation to ruin probabilities

u	0	0.25	0.5	0.75	1
$\psi(u)$	0.6111	0.5246	0.4547	0.3969	0.3479
$app(u)$	0.5508	0.4879	0.4322	0.3828	0.3391
Er	–9.87	–6.99	–4.97	–3.54	–2.54

u	1.25	1.5	1.75	2	2.25
$\psi(u)$	0.3059	0.2696	0.2379	0.2102	0.1858
$app(u)$	0.3003	0.2660	0.2357	0.2087	0.1849
Er	–1.82	–1.32	–0.95	–0.69	–0.50

This is equivalent to

$$cR(1 - G(u)) = (c - \lambda\mu)g(u) ,$$

and therefore has the solution

$$1 - G(u) = \exp\Bigl\{-\frac{cR}{c - \lambda\mu}u\Bigr\} .$$

The exponential distribution is thus the only one where the Cramér–Lundberg approximation becomes exact. ∎

5.7 Reinsurance and Ruin

5.7.1 Proportional Reinsurance

Recall that for proportional insurance the insurer covers $Y_i^I = \alpha Y_i$ of each claim, the reinsurer covers $Y_i^R = (1 - \alpha)Y_i$. Denote by c^I the insurer's premium rate. The insurer's adjustment coefficient is obtained from the equation

$$\lambda(M_{\alpha Y}(r) - 1) - c^I r = 0 .$$

The new moment generating function is

$$M_{\alpha Y}(r) = \mathbb{E}[e^{r\alpha Y_i}] = M_Y(\alpha r) .$$

Assume that both insurer and reinsurer use an expected value premium principle with the same safety loading. Then $c^I = \alpha c$ and we have to solve

$$\lambda(M_Y(\alpha r) - 1) - c\alpha r = 0 .$$

This is almost the original equation, hence $R = \alpha R^I$, where R^I is the adjustment coefficient under reinsurance. The new adjustment coefficient is larger, hence the risk has become smaller.

We assume now that there is an r_∞ such that $M_Y(r) < \infty$ if $r < r_\infty$ and $M_Y(r_\infty) = \infty$. Write $c = (1 + \xi)\lambda\mu$ for some $\xi > 0$. Suppose the reinsurer charges the premium $(1 + \vartheta)(1 - \alpha)\lambda\mu$. We now want to choose the retention level α so that the corresponding adjustment coefficient $R(\alpha)$ becomes maximal. In some sense, this means that the ruin probability is minimised for large initial capital, see also [85] or [116]. In order that the solution is not trivial, we assume $\vartheta > \xi$. $R(\alpha)$ is determined by the equation

$$\lambda[M_Y(\alpha R(\alpha)) - 1] - [\alpha(1 + \vartheta) - (\vartheta - \xi)]R(\alpha)\lambda\mu = 0 .$$

In order that the net profit condition is fulfilled, we need $\alpha(1+\vartheta)-(\vartheta-\xi)>\alpha$, or equivalently, $\alpha>1-\xi/\vartheta$. By the implicit function theorem, we have

$$\begin{aligned}0 = R'(\alpha)[\alpha M_Y'(\alpha R(\alpha)) - \{\alpha(1+\vartheta)-(\vartheta-\xi)\}\mu] + R(\alpha)M_Y'(\alpha R(\alpha))\\ - (1+\vartheta)R(\alpha)\mu\;.\end{aligned}$$

Thus, $R'(\alpha^*) = 0$ implies $M_Y'(\alpha^* R(\alpha^*)) = (1+\vartheta)\mu$. In particular, we get $M_Y'(\alpha^* R(\alpha^*)) - \{\alpha^*(1+\vartheta)-(\vartheta-\xi)\}\mu = \{(1-\alpha^*)(1+\vartheta)+(\vartheta-\xi)\}\mu > 0$. Differentiating a second time yields

$$\begin{aligned}0 = R''(\alpha)[\alpha M_Y'(\alpha R(\alpha)) - \{\alpha(1+\vartheta)-(\vartheta-\xi)\}\mu] + R'(\alpha)h(\alpha)\\ + R^2(\alpha)M_Y''(\alpha R(\alpha))\end{aligned}$$

for some function $h(\alpha)$. Because $R^2(\alpha)M_Y''(\alpha R(\alpha)) > 0$, we get that $R''(\alpha^*) < 0$. We see that $R(\alpha)$ has a unique maximum, and $R(\alpha)$ is increasing on $[1-\xi/\vartheta, \alpha^*\wedge 1]$ and decreasing on $[\alpha^*\wedge 1, 1]$.

We now write $r(\alpha) = \alpha R(\alpha)$. Then we have to solve

$$M_Y(r(\alpha)) - 1 - (1+\vartheta)r(\alpha)\mu + (\vartheta-\xi)r(\alpha)\mu/\alpha = 0\;.$$

We have seen above that $r^* = r(\alpha^*)$ is the unique value for which $M_Y'(r^*) = (1+\vartheta)\mu$. Note that such an $r^* < r_\infty$ exists under our assumptions.

$r(\alpha)$ is increasing on $[0, \alpha_0]$, where $\alpha_0 > \alpha^*$. We can therefore invert $r(\alpha)$, which we write as $\alpha(r)$. We get

$$\alpha(r) = \frac{(\vartheta-\xi)r\mu}{(1+\vartheta)r\mu - (M_Y(r)-1)}\;.$$

Thus the optimal retention level is

$$\alpha^* = \min\left\{\frac{(\vartheta-\xi)r^*\mu}{(1+\vartheta)r^*\mu - (M_Y(r^*)-1)}, 1\right\}.$$

If $\alpha(r^*) \le 1$ then $R(\alpha^*) = r^*/\alpha(r^*)$.

Example 5.12 Suppose that the claims are $\Gamma(\gamma,\alpha)$ distributed. Then $\mu = \gamma/\alpha$, $M_Y(r) = (\alpha/(\alpha-r))^\gamma$ and $M_Y'(r) = \gamma(\alpha/(\alpha-r))^{\gamma+1}/\alpha = (1+\vartheta)\gamma/\alpha$. Thus $r^* = \alpha(1-(1+\vartheta)^{-1/(\gamma+1)})$. Thus, after some simplifications,

$$\alpha^* = \min\left\{\frac{\gamma(\vartheta-\xi)[1-(1+\vartheta)^{-1/(\gamma+1)}]}{\gamma\vartheta + (\gamma+1)(1-(1+\vartheta)^{\gamma/(\gamma+1)})}, 1\right\}.$$

If $\alpha^* < 1$, then

$$R^* = \frac{\alpha\gamma(\vartheta - \xi)}{\gamma\vartheta + (\gamma+1)(1-(1+\vartheta)^{\gamma/(\gamma+1)})}\,.$$

Note that in general there is no closed formula for the adjustment coefficient for Gamma distributed claim sizes. However, for the maximal adjustment coefficient we obtain an explicit formula. ■

5.7.2 *Excess of Loss Reinsurance*

Under excess of loss reinsurance with retention level M the insurer has to pay $Y_i^I = \min\{Y_i, M\}$ of each claim. The adjustment coefficient is the strictly positive solution to the equation

$$\lambda\Bigl(\int_0^M e^{rx}\, dG(x) + e^{rM}(1 - G(M)) - 1\Bigr) - c^I r = 0\,.$$

There is no possibility of finding the solution to the problem without reinsurance. We have to solve the equation for every M separately. However, note that R^I exists in any case. Especially for heavy tailed distributions this shows that the risk has become much smaller. By the Cramér–Lundberg approximation the ruin probability decreases exponentially as the initial capital increases.

We will now show that for the insurer the excess of loss reinsurance is optimal. We assume that both insurer and reinsurer use an expected value principle.

Proposition 5.13 *Let all premia be computed via the expected value principle. Under all reinsurance forms acting on individual claims with premium rates c^I and c^R fixed, the excess of loss reinsurance maximises the insurer's adjustment coefficient.*

Proof Let $h(x)$ be an increasing function with $0 \le h(x) \le x$ for $x \ge 0$. We assume that the insurer pays $Y_i^I = h(Y_i)$. Let $h^*(x) = \min\{x, U\}$ be the excess of loss reinsurance. Because c^I is fixed, U can be determined from

$$\int_0^U y\, dG(y) + U(1 - G(U)) = \mathbb{E}[h^*(Y_i)] = \mathbb{E}[h(Y_i)] = \int_0^\infty h(y)\, dG(y)\,.$$

Because $e^z \ge 1 + z$, we obtain

$$e^{r(h(y)-h^*(y))} \ge 1 + r(h(y) - h^*(y))$$

and

$$e^{rh(y)} \ge e^{rh^*(y)}(1 + r(h(y) - h^*(y)))\,.$$

Thus

$$
\begin{aligned}
M_{h(Y)}(r) &= \int_0^\infty e^{rh(y)} \, dG(y) \ge \int_0^\infty e^{rh^*(y)}(1 + r(h(y) - h^*(y))) \, dG(y) \\
&= M_{h^*(Y)}(r) + r \int_0^\infty (h(y) - h^*(y)) \, e^{rh^*(y)} \, dG(y) \, .
\end{aligned}
$$

For $y \le U$ we have $h(y) \le y = h^*(y)$. We obtain for $r > 0$

$$
\begin{aligned}
&\int_0^\infty (h(y) - h^*(y))e^{rh^*(y)} \, dG(y) \\
&\quad = \int_0^U (h(y) - h^*(y))e^{rh^*(y)} \, dG(y) + \int_U^\infty (h(y) - h^*(y))e^{rh^*(y)} \, dG(y) \\
&\quad \ge \int_0^U (h(y) - h^*(y))e^{rU} \, dG(y) + \int_U^\infty (h(y) - h^*(y))e^{rU} \, dG(y) \\
&\quad = e^{rU} \int_0^\infty (h(y) - h^*(y)) \, dG(y) \\
&\quad = e^{rU} (\mathbb{E}[h(Y)] - \mathbb{E}[h^*(Y)]) = 0 \, .
\end{aligned}
$$

It follows that $M_{h(Y)}(r) \ge M_{h^*(Y)}(r)$ for $r > 0$. Since

$$
0 = \theta(R^I) = \lambda(M_{h(Y)}(R^I) - 1) - c^I R^I \ge \lambda(M_{h^*(Y)}(R^I) - 1) - c^I R^I = \theta^*(R^I)
$$

the assertion follows from the convexity of $\theta^*(r)$. □

Suppose now that we want to maximise the adjustment coefficient. Suppose again that the safety loading of the insurer is ξ and the safety loading of the reinsurer is $\vartheta > \xi$. Recall that

$$
\mathbb{E}[e^{r(Y \wedge M)} - 1] = \int_0^M r e^{ry}(1 - G(y)) \, dy
$$

and

$$
\mathbb{E}[\max\{Y - M, 0\}] = \int_M^\infty (1 - G(y)) \, dy \, .
$$

The net profit condition becomes, after dividing by λ,

$$
(1 + \xi) \int_0^M (1 - G(y)) \, dy - (\vartheta - \xi) \int_M^\infty (1 - G(y)) \, dy > \int_0^M (1 - G(y)) \, dy \, ,
$$

or equivalently

$$
\xi\mu > \vartheta \int_M^\infty (1 - G(y)) \, dy \, .
$$

For the adjustment coefficient we have to find the solution to

$$\lambda \int_0^M r(e^{ry} - (1+\xi))(1 - G(y))\, dy + (\vartheta - \xi)\lambda r \int_M^\infty (1 - G(y))\, dy = 0\,.$$

Since $R \neq 0$, we look for the solution to

$$\int_0^M (e^{ry} - (1+\xi))(1 - G(y))\, dy + (\vartheta - \xi) \int_M^\infty (1 - G(y))\, dy = 0\,.$$

Taking the derivative, we get from the implicit function theorem

$$R'(M) \int_0^M y e^{R(M)y}(1 - G(y))\, dy + [e^{R(M)M} - (1+\vartheta)](1 - G(M)) = 0\,.$$

If $R'(M) = 0$, then $e^{R(M)M} = 1 + \vartheta$. We conclude that it is not optimal not to take reinsurance. Indeed, as $M \to \infty$ the adjustment coefficient $R(M)$ tends to the adjustment coefficient of the model without reinsurance, and therefore $R(M)M \to \infty$. If the adjustment coefficient does not exist but exponential moments exist, then the ruin probability decreases exponentially at the rate r_∞, and also $r_\infty M \to \infty$ as $M \to \infty$. Finally, if no exponential moments exist, the adjustment coefficient exists after reinsurance.

Suppose for the moment that $G(y)$ is differentiable. Then the second derivative gives

$$\begin{aligned} R''(M) \int_0^M y e^{R(M)y}(1 - G(y))\, dy + R'(M)h(M) \\ + R(M)e^{R(M)M}(1 - G(M)) - [e^{R(M)M} - (1+\vartheta)]G'(M) = 0 \end{aligned}$$

for some function $h(M)$. At a point where $R'(M) = 0$, we have

$$R''(M) \int_0^M y e^{R(M)y}(1 - G(y))\, dy + R(M)e^{R(M)M}(1 - G(M)) = 0\,,$$

where we used that $e^{R(M)M} = 1 + \vartheta$. We conclude that $R''(M) < 0$. This shows that $R(M)$ has a unique local maximum, and hence is unimodal. For $M < M^*$, the optimal retention level, we have $R'(M) > 0$, and thus $e^{R(M)M} < 1+\vartheta$. For $M > M^*$, we have $R'(M) < 0$, and therefore $e^{R(M)M} > 1 + \vartheta$. The solution is thus found by calculating $R(M)M$, and we find the unique value for which $R(M)M = \log(1+\vartheta)$.

We consider now the portfolio of the reinsurer. What is the claim number process of the claims the reinsurer is involved in? Because the claim amounts are independent of the claim arrival process we delete independently points from the Poisson process with probability $G(M)$ and do not delete them with probability $1 - G(M)$. By Proposition C.5 this process is a Poisson process with rate $\lambda(1 - G(M))$. Because the claim sizes are iid. and independent of the claim arrival process the surplus of the reinsurer is a Cramér–Lundberg process with intensity $\lambda(1 - G(M))$ and claim size distribution

$$\tilde{G}(x) = \mathbb{P}[Y_i - M \le x \mid Y_i > M] = \frac{G(M + x) - G(M)}{1 - G(M)}\;.$$

5.8 The Severity of Ruin, the Capital Prior to Ruin and the Distribution of $\inf\{C_t : t \ge 0\}$

For an insurance company ruin is not so dramatic if $-C_\tau$ is small, but it could ruin the whole company if $-C_\tau$ is very large. So we are interested in the distribution of $-C_\tau$ if ruin occurs. The random variable $-C_\tau$ is called the *severity of ruin*. We further want to know how ruin happens. Thus, we also consider the capital prior to ruin $C_{\tau-}$. Let

$$\psi_{x,y}(u) = \mathbb{P}[\tau < \infty, C_\tau < -x, C_{\tau-} > y]\;.$$

We proceed as in Sect. 5.3. For h small

$$\psi_{x,y}(u) = \mathrm{e}^{-\lambda h}\psi_{x,y}(u + ch) + \int_0^h \Bigl[\int_0^{u+ct} \psi_{x,y}(u + ct - z)\,\mathrm{d}G(z) + \mathbb{I}_{u+ct>y}(1 - G(u + ct + x))\Bigr]\lambda \mathrm{e}^{-\lambda t}\,\mathrm{d}t\;.$$

It follows that $\psi_{x,y}(u)$ is right continuous. We obtain the integro-differential equation

$$c\psi_x'(u) = \lambda\Bigl[\psi_{x,y}(u) - \int_0^u \psi_{x,y}(u - z)\,\mathrm{d}G(z) - \mathbb{I}_{u\ge y}(1 - G(u + x))\Bigr]\,.$$

Replacing u by $u - ch$ gives the derivative from the left. However, $\psi_{x,y}(u)$ is not differentiable at points where $G(u)$ jumps or at $u = y$. Integration of the integro-differential equation yields

$$\begin{aligned}
&\frac{c}{\lambda}(\psi_{x,y}(u) - \psi_{x,y}(0))\\
&\quad= \int_0^u \psi_{x,y}(u - t)(1 - G(t))\,\mathrm{d}t - \int_{u\wedge y}^u (1 - G(t + x))\,\mathrm{d}t\\
&\quad= \int_0^u \psi_{x,y}(u - t)(1 - G(t))\,\mathrm{d}t - \int_{(x+u)\wedge(x+y)}^{x+u} (1 - G(t))\,\mathrm{d}t\;.
\end{aligned}$$

Because $\psi_{x,y}(u) \le \psi(u)$ we can see that $\psi_{x,y}(u) \to 0$ as $u \to \infty$. By the bounded convergence theorem we obtain

$$-\frac{c}{\lambda}\psi_{x,y}(0) = -\int_{x+y}^{\infty} (1 - G(t))\,\mathrm{d}t$$

and

$$\mathbb{P}[C_\tau < -x, C_{\tau-} > y \mid \tau < \infty, C_0 = 0] = \frac{\dfrac{\lambda}{c}\displaystyle\int_{x+y}^{\infty} (1 - G(t))\,\mathrm{d}t}{\dfrac{\lambda\mu}{c}} = \frac{1}{\mu}\int_{x+y}^{\infty} (1 - G(t))\,\mathrm{d}t\,. \tag{5.9}$$

If we choose $y = u = 0$, we get that $-C_\tau$ has the conditional distribution $B(x) = \mu^{-1}\int_0^x (1 - G(t))\,\mathrm{d}t$ given that ruin occurs.

Remark 5.14 In order to get an equation for the ruin probability we can consider the first time point τ_1 where the surplus is below the initial capital. At this point, by the strong Markov property, a new Cramér–Lundberg process starts. We get three possibilities:

- The process never falls below the initial capital.
- $\tau_1 < \infty$ but $C_{\tau_1} \ge 0$.
- Ruin occurs at τ_1.

Thus we get

$$\psi(u) = \Bigl(1 - \frac{\lambda\mu}{c}\Bigr)0 + \frac{\lambda}{c}\int_0^u \psi(u - y)(1 - G(y))\,\mathrm{d}y + \frac{\lambda}{c}\int_u^{\infty} 1\,(1 - G(y))\,\mathrm{d}y\,.$$

This is Eq. (5.2). Thus we now have a natural interpretation of (5.2). ■

Let $\tau_0 = 0$ and $\tau_i = \inf\{t > \tau_{i-1} : C_t < C_{\tau_{i-1}}\}$, called the *ladder times*, and define $L_i = C_{\tau_{i-1}} - C_{\tau_i}$, called the *ladder heights*. Note that L_i is only defined if $\tau_i < \infty$. By Lemma 5.1 we find $\mathbb{P}[\tau_i < \infty \mid \tau_{i-1} < \infty] = \lambda\mu/c$. Let $K = \sup\{i \in \mathbb{N} : \tau_i < \infty\}$ be the number of ladder epochs. We have just seen that $K \sim \mathrm{NB}(1, 1 - \lambda\mu/c)$ and that, given K, the random variables $(L_i : i \le K)$ are iid. and absolutely continuous with density $(1 - G(x))/\mu$. We only have to condition on K because L_i is not defined for $i > K$. If we assume that all $(L_i : i \ge 1)$ have the same distribution, then we can drop the conditioning on K and (L_i) is independent of K. Then

$$\inf\{C_t : t \ge 0\} = u - \sum_{i=1}^{K} L_i$$

and

$$\mathbb{P}[\tau < \infty] = \mathbb{P}[\inf\{C_t : t \geq 0\} < 0] = \mathbb{P}\Big[\sum_{i=1}^{K} L_i > u\Big].$$

We can use Panjer recursion to approximate $\psi(u)$ by using an appropriate discretisation.

A formula that is useful for theoretical considerations, but too messy to use for the computation of $\psi(u)$, is the Pollaczek–Khintchine formula

$$\begin{aligned}\psi(u) &= \mathbb{P}\Big[\sum_{i=1}^{K} L_i > u\Big] = \sum_{n=1}^{\infty} \mathbb{P}\Big[\sum_{i=1}^{n} L_i > u\Big]\mathbb{P}[K = n] \\ &= \Big(1 - \frac{\lambda\mu}{c}\Big)\sum_{n=1}^{\infty}\Big(\frac{\lambda\mu}{c}\Big)^n (1 - B^{*n}(u)) .\end{aligned} \tag{5.10}$$

5.9 The Laplace Transform of ψ

Definition 5.15 Let f be a real function on $[0, \infty)$. The transform

$$\hat{f}(s) := \int_0^{\infty} \mathrm{e}^{-sx} f(x)\, \mathrm{d}x \qquad (s \in \mathbb{R})$$

is called the *Laplace transform* of f.

If X is an absolutely continuous positive random variable with density f then $\hat{f}(s) = M_X(-s)$. The Laplace transform has the following properties.

Lemma 5.16

(i) If $f(x) \geq 0$ a.e.

$$\hat{f}(s_1) \leq \hat{f}(s_2) \iff s_1 \geq s_2.$$

(ii)

$$\widehat{|f|}(s_1) < \infty \implies \widehat{|f|}(s) < \infty \text{ for all } s \geq s_1.$$

(iii)

$$\widehat{f'}(s) = \int_0^{\infty} \mathrm{e}^{-sx} f'(x)\, \mathrm{d}x = s\hat{f}(s) - f(0)$$

provided $f'(x)$ exists a.e. and $\widehat{|f|}(s) < \infty$.

(iv)

$$\lim_{s\to\infty} s\hat{f}(s) = \lim_{x\to 0} f(x)$$

provided $f'(x)$ *exists a.e.,* $\lim_{x\to 0} f(x)$ *exists and* $\widehat{|f|}(s) < \infty$ *for large enough* s.

(v)

$$\lim_{s\downarrow 0} s\hat{f}(s) = \lim_{x\to\infty} f(x)$$

provided $f'(x)$ *exists a.e.,* $\lim_{x\to\infty} f(x)$ *exists and* $\widehat{|f|}(s) < \infty$ *for some* $s > 0$.

(vi)

$$\hat{f}(s) = \hat{g}(s) \text{ on } (s_0, s_1) \quad\Longrightarrow\quad f(x) = g(x) \text{ Lebesgue a.e. } \quad \forall x \in [0,\infty).$$

□

We want to find the Laplace transform of $\delta(u)$. We multiply (5.1) by e^{-su} and then integrate over u. Let $s > 0$. Then

$$c\int_0^\infty \delta'(u)\mathrm{e}^{-su}\,\mathrm{d}u = \lambda\int_0^\infty \delta(u)\mathrm{e}^{-su}\,\mathrm{d}u - \lambda\int_0^\infty\int_0^u \delta(u-y)\,\mathrm{d}G(y)\mathrm{e}^{-su}\,\mathrm{d}u.$$

We have to determine the last integral.

$$\begin{aligned}\int_0^\infty\int_0^u \delta(u-y)\,\mathrm{d}G(y)\mathrm{e}^{-su}\,\mathrm{d}u &= \int_0^\infty\int_y^\infty \delta(u-y)\mathrm{e}^{-su}\,\mathrm{d}u\,\mathrm{d}G(y)\\ &= \int_0^\infty\int_0^\infty \delta(u)\mathrm{e}^{-s(u+y)}\,\mathrm{d}u\,\mathrm{d}G(y) = \hat{\delta}(s)M_Y(-s)\,.\end{aligned}$$

Thus we get the equation

$$c(s\hat{\delta}(s) - \delta(0)) = \lambda\hat{\delta}(s)(1 - M_Y(-s))\,,$$

which has the solution

$$\hat{\delta}(s) = \frac{c\delta(0)}{cs - \lambda(1 - M_Y(-s))} = \frac{c - \lambda\mu}{cs - \lambda(1 - M_Y(-s))}\,.$$

The Laplace transform of ψ can easily be found as

$$\hat{\psi}(s) = \int_0^\infty (1 - \delta(u))\mathrm{e}^{-su}\,\mathrm{d}u = \frac{1}{s} - \hat{\delta}(s)\,.$$

Example 5.17 For exponentially distributed claims we obtain

$$\hat{\delta}(s) = \frac{c - \lambda/\alpha}{cs - \lambda(1 - \alpha/(\alpha + s))} = \frac{c - \lambda/\alpha}{s(c - \lambda/(\alpha + s))} = \frac{(c - \lambda/\alpha)(\alpha + s)}{s(c(\alpha + s) - \lambda)}$$

and

$$\begin{aligned}\hat{\psi}(s) &= \frac{1}{s} - \frac{(c - \lambda/\alpha)(\alpha + s)}{s(c(\alpha + s) - \lambda)} = \frac{c(\alpha + s) - \lambda - (c - \lambda/\alpha)(\alpha + s)}{s(c(\alpha + s) - \lambda)} \\ &= \frac{\lambda}{\alpha}\frac{1}{c(\alpha + s) - \lambda} = \frac{\lambda}{\alpha c}\frac{1}{\alpha - \lambda/c + s}.\end{aligned}$$

By comparison with the moment generating function of the exponential distribution we recognise that

$$\psi(u) = \frac{\lambda}{\alpha c}\mathrm{e}^{-(\alpha - \lambda/c)u}.$$

■

Example 5.18 Let $G(x) = 1 - p\mathrm{e}^{-\alpha x} - (1 - p)\mathrm{e}^{-\beta x}$ where $0 < \alpha < \beta$ and $0 < p \le \beta(\beta - \alpha)^{-1}$. For the Laplace transform of δ we get

$$\begin{aligned}\hat{\delta}(s) &= \frac{c - \lambda p/\alpha - \lambda(1 - p)/\beta}{cs - \lambda(1 - \alpha p/(\alpha + s) - \beta(1 - p)/(\beta + s))} \\ &= \frac{c - \lambda p/\alpha - \lambda(1 - p)/\beta}{s(c - \lambda p/(\alpha + s) - \lambda(1 - p)/(\beta + s))}\end{aligned}$$

and for the Laplace transform of ψ

$$\begin{aligned}\hat{\psi}(s) &= \frac{c - \lambda p/(\alpha + s) - \lambda(1 - p)/(\beta + s) - c + \lambda p/\alpha + \lambda(1 - p)/\beta}{s(c - \lambda p/(\alpha + s) - \lambda(1 - p)/(\beta + s))} \\ &= \lambda\frac{sp/(\alpha(\alpha + s)) + s(1 - p)/(\beta(\beta + s))}{s(c - \lambda p/(\alpha + s) - \lambda(1 - p)/(\beta + s))} \\ &= \lambda\frac{p(\beta + s)/\alpha + (1 - p)(\alpha + s)/\beta}{c(\alpha + s)(\beta + s) - \lambda p(\beta + s) - \lambda(1 - p)(\alpha + s)} \\ &= \lambda\frac{p(\beta + s)/\alpha + (1 - p)(\alpha + s)/\beta}{cs^2 + ((\alpha + \beta)c - \lambda)s + \alpha\beta c - \lambda((1 - p)\alpha + p\beta)}.\end{aligned}$$

The denominator can be written as $c(s+R)(s+\bar{R})$ where R and $\bar{R}$ are the two solutions to (5.6) with $R < \bar{R}$. The Laplace transform of ψ can be written in the form

$$\hat{\psi}(s) = \frac{A}{R + s} + \frac{B}{\bar{R} + s}$$

for some constants A and B. Hence

$$\psi(u) = A\mathrm{e}^{-Ru} + B\mathrm{e}^{-\bar{R}u}\,.$$

A must be the constant appearing in the Cramér–Lundberg approximation. The constant B can be found from $\psi(0)$. ∎

Recall that τ_1 is the first ladder epoch. On the set $\{\tau_1 < \infty\}$ the random variable $\sup\{u - C_t : t \ge 0\}$ is absolutely continuous with distribution function

$$\mathbb{P}[\sup\{u - C_t : t \ge 0\} \le x \mid \tau_1 < \infty] = 1 - \frac{c}{\lambda\mu}\psi(x)\,.$$

Let Z be a random variable with the above distribution. Its moment generating function is

$$\begin{aligned} M_Z(r) &= \int_0^\infty \mathrm{e}^{ru}\frac{c}{\lambda\mu}\delta'(u)\,\mathrm{d}u = \frac{c}{\lambda\mu}\Bigl(-r\frac{c-\lambda\mu}{-cr-\lambda(1-M_Y(r))} - \Bigl(1-\frac{\lambda\mu}{c}\Bigr)\Bigr) \\ &= 1 - \frac{c}{\lambda\mu} + \frac{c(c-\lambda\mu)}{\lambda\mu}\,\frac{r}{cr-\lambda(M_Y(r)-1)}\,. \end{aligned}$$

We will later need the first two moments of the above distribution function. Assume that $\mu_2 < \infty$. The first derivative of the moment generating function is

$$\begin{aligned} M_Z'(r) &= \frac{c(c-\lambda\mu)}{\lambda\mu}\,\frac{cr-\lambda(M_Y(r)-1)-r(c-\lambda M_Y'(r))}{(cr-\lambda(M_Y(r)-1))^2} \\ &= \frac{c(c-\lambda\mu)}{\mu}\,\frac{rM_Y'(r)-(M_Y(r)-1)}{(cr-\lambda(M_Y(r)-1))^2}\,. \end{aligned}$$

Note that

$$\lim_{r\to 0}\frac{1}{r}(cr-\lambda(M_Y(r)-1)) = c-\lambda\mu\,.$$

We find

$$\lim_{r\to 0}\frac{rM_Y'(r)-(M_Y(r)-1)}{r^2} = \lim_{r\to 0}\frac{M_Y'(r)+rM_Y''(r)-M_Y'(r)}{2r} = \frac{\mu_2}{2}$$

and thus

$$\mathbb{E}[Z] = \frac{c(c-\lambda\mu)}{\mu}\,\frac{\mu_2}{2(c-\lambda\mu)^2} = \frac{c\mu_2}{2\mu(c-\lambda\mu)}\,. \tag{5.11}$$

Assume now that $\mu_3 < \infty$. The second derivative of $M_Z(r)$ is

$$M_Z''(r) = \frac{c(c-\lambda\mu)}{\mu} \frac{1}{(cr-\lambda(M_Y(r)-1))^3} \Bigl(rM_Y''(r)(cr-\lambda(M_Y(r)-1)) \\ - 2(rM_Y'(r)-(M_Y(r)-1))(c-\lambda M_Y'(r))\Bigr)\;.$$

For the limit to 0 we find

$$\begin{aligned}
&\lim_{r\to 0}\frac{1}{r^3}\Bigl(rM_Y''(r)(cr-\lambda(M_Y(r)-1)) \\
&\qquad\qquad - 2(rM_Y'(r)-(M_Y(r)-1))(c-\lambda M_Y'(r))\Bigr) \\
&= \lim_{r\to 0}\frac{1}{3r^2}\Bigl((M_Y''(r)+rM_Y'''(r))(cr-\lambda(M_Y(r)-1)) \\
&\qquad + rM_Y''(r)(c-\lambda M_Y'(r)) - 2rM_Y''(r)(c-\lambda M_Y'(r)) \\
&\qquad + 2(rM_Y'(r)-(M_Y(r)-1))\lambda M_Y''(r)\Bigr) \\
&= \frac{\mu_3}{3}(c-\lambda\mu) + \lambda\mu_2 \lim_{r\to 0}\frac{rM_Y'(r)-(M_Y(r)-1)}{r^2} \\
&= \frac{\mu_3}{3}(c-\lambda\mu) + \frac{\lambda}{2}\mu_2^2\;.
\end{aligned}$$

Thus the second moment of Z becomes

$$\mathbb{E}[Z^2] = \frac{c(c-\lambda\mu)}{\mu}\frac{\mu_3(c-\lambda\mu)/3+\lambda\mu_2^2/2}{(c-\lambda\mu)^3} = \frac{c}{\mu}\Bigl(\frac{\mu_3}{3(c-\lambda\mu)} + \frac{\lambda\mu_2^2}{2(c-\lambda\mu)^2}\Bigr)\;. \tag{5.12}$$

5.10 Approximations to ψ

5.10.1 Diffusion Approximations

Diffusion approximations are based on the following

Proposition 5.19 *Let $\{C_t^{(n)}\}$ be a sequence of Cramér–Lundberg processes with initial capital $u^{(n)} = u$, claim arrival intensities $\lambda^{(n)} = \lambda n$, claim size distributions $G^{(n)}(x) = G(x\sqrt{n})$ and premium rates*

$$c^{(n)} = \Bigl(1 + \frac{c-\lambda\mu}{\lambda\mu\sqrt{n}}\Bigr)\lambda^{(n)}\mu^{(n)} = c + (\sqrt{n}-1)\lambda\mu\;.$$

Let $\mu = \int_0^\infty y \, \mathrm{d}G(y)$ and assume that $\mu_2 = \int_0^\infty y^2 \, \mathrm{d}G(y) < \infty$. Then

$$\{C_t^{(n)}\} \xrightarrow{\mathrm{d}} \{u + W_t\}$$

in distribution in the topology of uniform convergence on finite intervals where $\{W_t\}$ is a $(c - \lambda\mu, \lambda\mu_2)$-Brownian motion.

Proof See [86] or [73]. □

Intuitively we let the number of claims in a unit time interval go to infinity and make the claim sizes smaller in such a way that the distribution of $C_1^{(n)} - u$ tends to a normal distribution and $\mathbb{E}[C_1^{(n)} - u] = c - \lambda\mu$. Let $\tau^{(n)}$ denote the ruin time of $\{C_t^{(n)}\}$ and $\tau = \inf\{t \geq 0 : u + W_t < 0\}$ the ruin probability of the Brownian motion. Then

Proposition 5.20 *Let $(C_t^{(n)})$ and (W_t) be as above. Then*

$$\lim_{n\to\infty} \mathbb{P}[\tau^{(n)} \leq t] = \mathbb{P}[\tau \leq t]$$

and

$$\lim_{n\to\infty} \mathbb{P}[\tau^{(n)} < \infty] = \mathbb{P}[\tau < \infty]\,.$$

Proof The result for a finite time horizon is a special case of [138, Thm. 9], see also [86] or [73]. The result for the infinite time horizon can be found in [113]. □

The idea of the diffusion approximation is to approximate $\mathbb{P}[\tau^{(1)} \leq t]$ by $\mathbb{P}[\tau \leq t]$ and $\mathbb{P}[\tau^{(1)} < \infty]$ by $\mathbb{P}[\tau < \infty]$. Thus we need the ruin probabilities of the Brownian motion.

Lemma 5.21 *Let $\{W_t\}$ be a (m, η^2)-Brownian motion with $m > 0$ and $\tau = \inf\{t \geq 0 : u + W_t < 0\}$. Then*

$$\mathbb{P}[\tau < \infty] = \mathrm{e}^{-2um/\eta^2}$$

and

$$\mathbb{P}[\tau \leq t] = 1 - \Phi\Bigl(\frac{mt + u}{\eta\sqrt{t}}\Bigr) + \mathrm{e}^{-2um/\eta^2}\,\Phi\Bigl(\frac{mt - u}{\eta\sqrt{t}}\Bigr)\,.$$

Proof By Lemma D.3 the process

$$\exp\Bigl\{-\frac{2m(u + W_t)}{\eta^2}\Bigr\}$$

is a martingale. By the stopping theorem

$$\exp\Bigl\{-\frac{2m(u + W_{\tau\wedge t})}{\eta^2}\Bigr\}$$

is a positive bounded martingale. Thus, because $\lim_{t\to\infty} W_t = \infty$,

$$\exp\Bigl\{-\frac{2um}{\eta^2}\Bigr\} = \mathbb{E}\Bigl[\exp\Bigl\{-\frac{2m(u+W_\tau)}{\eta^2}\Bigr\}\Bigr] = \mathbb{P}[\tau < \infty]\,.$$

It is easy to see that $\{sW_{1/s} - m\}$ is a $(0,\eta^2)$-Brownian motion (see [89, p. 351]) and thus $\{s(u + W_{1/s}) - m\}$ is a (u,η^2)-Brownian motion. Denote the latter process by $\{\tilde{W}_s\}$. Then

$$\begin{aligned}
\mathbb{P}[\tau \le t] &= \mathbb{P}[\inf\{u + W_s : 0 < s \le t\} < 0] \\
&= \mathbb{P}[\inf\{s(u + W_{1/s}) : s \ge 1/t\} < 0] \\
&= \mathbb{E}[\mathbb{P}[\inf\{m + \tilde{W}_s : s \ge 1/t\} < 0 \mid \tilde{W}_{1/t}]] \\
&= \int_{-\infty}^{-m} \frac{1}{\sqrt{2\pi\eta^2/t}} \mathrm{e}^{-\frac{(y-u/t)^2}{2\eta^2/t}}\,\mathrm{d}y \\
&\qquad + \int_{-m}^{\infty} \mathrm{e}^{-\frac{2u(y+m)}{\eta^2}} \frac{1}{\sqrt{2\pi\eta^2/t}} \mathrm{e}^{-\frac{(y-u/t)^2}{2\eta^2/t}}\,\mathrm{d}y \\
&= \Phi\Bigl(\frac{-m - u/t}{\eta/\sqrt{t}}\Bigr) + \mathrm{e}^{-\frac{2um}{\eta^2}} \int_{-m}^{\infty} \frac{1}{\sqrt{2\pi\eta^2/t}} \mathrm{e}^{-\frac{(y+u/t)^2}{2\eta^2/t}}\,\mathrm{d}y \\
&= 1 - \Phi\Bigl(\frac{mt + u}{\eta\sqrt{t}}\Bigr) + \mathrm{e}^{-\frac{2um}{\eta^2}} \Phi\Bigl(\frac{mt - u}{\eta\sqrt{t}}\Bigr)\,.
\end{aligned}$$

□

Diffusion approximations only work well if $c/(\lambda\mu)$ is close to 1. There also exist corrected diffusion approximations which are much better, see [122] or [9].

Example 5.22 Let $c = \lambda = 1$ and $G(x) = 1 - \frac{1}{3}(\mathrm{e}^{-x} + \mathrm{e}^{-2x} + \mathrm{e}^{-3x})$. We find $c - \lambda\mu = 7/18$ and $\lambda\mu_2 = 49/54$. This leads to the diffusion approximation $\psi(u) \approx \exp\{-6u/7\}$. Table 5.2 shows exact values ($\psi(u)$), the diffusion approximation (DA) and the relative error multiplied by 100 (Er). Here we have $c/(\lambda\mu) = 18/11 = 1.63636$ is not close to one. This is also indicated by the figures. ■

5.10.2 The deVylder Approximation

In the case of exponentially distributed claim amounts we know the ruin probabilities explicitly. The idea of the deVylder approximation is to replace $\{C_t\}$ by $\{\tilde{C}_t\}$ where $\{\tilde{C}_t\}$ has exponentially distributed claim amounts and

$$\mathbb{E}[(C_t - u)^k] = \mathbb{E}[(\tilde{C}_t - u)^k] \qquad \text{for } k = 1, 2, 3.$$

Table 5.2 Diffusion approximation to ruin probabilities

u	0	0.25	0.5	0.75	1
$\psi(u)$	0.6111	0.5246	0.4547	0.3969	0.3479
DA	1.0000	0.8071	0.6514	0.5258	0.4244
Er	63.64	53.87	43.26	32.49	21.98

u	1.25	1.5	1.75	2	2.25
$\psi(u)$	0.3059	0.2696	0.2379	0.2102	0.1858
DA	0.3425	0.2765	0.2231	0.1801	0.1454
Er	11.96	2.54	–6.22	–14.32	–21.78

The first three (centralised) moments are

$$\mathbb{E}[C_t - u] = (c - \lambda\mu)t = \Bigl(\tilde{c} - \frac{\tilde{\lambda}}{\tilde{\alpha}}\Bigr)t\,,$$

$$\mathrm{Var}[C_t] = \mathrm{Var}[u + ct - C_t] = \lambda\mu_2 t = \frac{2\tilde{\lambda}}{\tilde{\alpha}^2}t$$

and

$$\mathbb{E}[(C_t - \mathbb{E}[C_t])^3] = -\mathbb{E}[(u + ct - C_t - \mathbb{E}[u + ct - C_t])^3] = -\lambda\mu_3 t = -\frac{6\tilde{\lambda}}{\tilde{\alpha}^3}t\,.$$

The parameters of the approximation are

$$\tilde{\alpha} = \frac{3\mu_2}{\mu_3}\,,$$

$$\tilde{\lambda} = \frac{\lambda\mu_2\tilde{\alpha}^2}{2} = \frac{9\mu_2^3}{2\mu_3^2}\lambda$$

and

$$\tilde{c} = c - \lambda\mu + \frac{\tilde{\lambda}}{\tilde{\alpha}} = c - \lambda\mu + \frac{3\mu_2^2}{2\mu_3}\lambda\,.$$

Thus the approximation to the probability of ultimate ruin is

$$\psi(u) \approx \frac{\tilde{\lambda}}{\tilde{\alpha}\tilde{c}}\mathrm{e}^{-\left(\tilde{\alpha} - \frac{\tilde{\lambda}}{\tilde{c}}\right)u}\,.$$

Table 5.3 DeVylder approximation to ruin probabilities

u	0	0.25	0.5	0.75	1
$\psi(u)$	0.6111	0.5246	0.4547	0.3969	0.3479
DV	0.5774	0.5102	0.4509	0.3984	0.3520
Er	-5.51	-2.73	-0.86	0.38	1.18

u	1.25	1.5	1.75	2	2.25
$\psi(u)$	0.3059	0.2696	0.2379	0.2102	0.1858
DV	0.3110	0.2748	0.2429	0.2146	0.1896
Er	1.67	1.95	2.07	2.09	2.03

There is also a formula for the probability of ruin within finite time. Let $\eta = \sqrt{\tilde{\lambda}/(\tilde{\alpha}\tilde{c})}$. Then

$$\psi(u,t) \approx \frac{\tilde{\lambda}}{\tilde{\alpha}\tilde{c}} e^{-(\tilde{\alpha}-\tilde{\lambda}/\tilde{c})u} - \frac{1}{\pi}\int_0^\pi f(x)\,dx \tag{5.13}$$

where

$$f(x) = \eta \frac{\exp\{2\eta\tilde{\alpha}\tilde{c}t\cos x - (\tilde{\alpha}\tilde{c}+\tilde{\lambda})t + \tilde{\alpha}u(\eta\cos x - 1)\}}{1+\eta^2 - 2\eta\cos x} \cdot (\cos(\tilde{\alpha}u\eta\sin x) - \cos(\tilde{\alpha}u\eta\sin x + 2x))\,.$$

A numerical investigation shows that the approximation is quite accurate.

Example 5.23 Let $c = \lambda = 1$ and $G(x) = 1 - \frac{1}{3}(e^{-x} + e^{-2x} + e^{-3x})$. In addition to the previously calculated values we also need $\mu_3 = 251/108$. The approximation parameters are $\tilde{\alpha} = 1.17131$, $\tilde{\lambda} = 0.622472$ and $\tilde{c} = 0.920319$. This leads to the approximation $\psi(u) \approx 0.577441e^{-0.494949u}$. It turns out that the approximation works well (Table 5.3). ■

5.10.3 The Beekman–Bowers Approximation

Recall from (5.11) and (5.12) that

$$F(u) = 1 - \frac{c}{\lambda\mu}\psi(u)$$

is a distribution function and that

$$\int_0^\infty z\,dF(z) = \frac{c\mu_2}{2\mu(c-\lambda\mu)}$$

and that
$$\int_0^\infty z^2 \,\mathrm{d}F(z) = \frac{c}{\mu}\Bigl(\frac{\mu_3}{3(c-\lambda\mu)} + \frac{\lambda\mu_2^2}{2(c-\lambda\mu)^2}\Bigr).$$

The idea is to approximate the distribution function F by the distribution function $\tilde{F}(u)$ of a $\Gamma(\gamma, \alpha)$ distributed random variable such that the first two moments coincide. Thus the parameters γ and α have to satisfy

$$\begin{aligned}\frac{\gamma}{\alpha} &= \frac{c\mu_2}{2\mu(c-\lambda\mu)},\\ \frac{\gamma(\gamma+1)}{\alpha^2} &= \frac{c}{\mu}\Bigl(\frac{\mu_3}{3(c-\lambda\mu)} + \frac{\lambda\mu_2^2}{2(c-\lambda\mu)^2}\Bigr).\end{aligned}$$

The Beekman–Bowers approximation to the ruin probability is

$$\psi(u) = \frac{\lambda\mu}{c}(1-F(u)) \approx \frac{\lambda\mu}{c}(1-\tilde{F}(u)).$$

Remark 5.24 If $2\gamma \in \mathbb{N}$ then $2\alpha Z \sim \chi^2_{2\gamma}$ is χ^2 distributed. ∎

Example 5.25 Let $c = \lambda = 1$ and $G(x) = 1 - \frac{1}{3}(\mathrm{e}^{-x} + \mathrm{e}^{-2x} + \mathrm{e}^{-3x})$. For the Beekman–Bowers approximation we have to solve the equations

$$\frac{\gamma}{\alpha} = 1.90909, \qquad \frac{\gamma(\gamma+1)}{\alpha^2} = 7.71429,$$

which yields the parameters $\gamma = 0.895561$ and $\alpha = 0.469104$. From this the Beekman–Bowers approximation can be obtained. Here we have $2\gamma = 1.79112$, which is not close to an integer. Anyway, one can interpolate between the χ^2_1 and the χ^2_2 distribution function to get the approximation

$$0.20888\chi^2_1(2\alpha u) + 0.79112\chi^2_2(2\alpha u)$$

to $1 - c/(\lambda\mu)\,\psi(u)$. Table 5.4 shows the exact values ($\psi(u)$), the Beekman–Bowers approximation (BB1) and the approximation obtained by interpolating the χ^2 distributions (BB2). The relative errors (Er) are given in percent. One can clearly see that all the approximations work well. ∎

5.11 Subexponential Claim Size Distributions

Let us now consider subexponential claim size distributions. In this case the Lundberg exponent does not exist (Lemma F.3).

Table 5.4 Beekman–Bowers approximation to ruin probabilities

u	0	0.25	0.5	0.75	1
$\psi(u)$	0.6111	0.5246	0.4547	0.3969	0.3479
BB1	0.6111	0.5227	0.4553	0.3985	0.3498
Er	0.00	–0.35	0.12	0.42	0.54
BB2	0.6111	0.5105	0.4456	0.3914	0.3450
Er	0.00	–2.68	–2.02	–1.38	–0.83

u	1.25	1.5	1.75	2	2.25
$\psi(u)$	0.3059	0.2696	0.2379	0.2102	0.1858
BB1	0.3076	0.2709	0.2387	0.2106	0.1859
Er	0.54	0.47	0.34	0.19	0.04
BB2	0.3046	0.2693	0.2383	0.2110	0.1869
Er	–0.42	–0.11	0.18	0.40	0.59

Theorem 5.26 *Assume that the ladder height distribution*

$$\frac{1}{\mu}\int_0^x (1 - G(y))\, dy$$

is subexponential. Then

$$\lim_{u\to\infty} \frac{\psi(u)}{\int_u^\infty (1 - G(y))\, dy} = \frac{\lambda}{c - \lambda\mu}\,.$$

Remark 5.27 Recall that the probability that ruin occurs at the first ladder time given there is a first ladder epoch is

$$\frac{1}{\mu}\int_u^\infty (1 - G(y))\, dy\,.$$

Hence the ruin probability is asymptotically $(c - \lambda\mu)^{-1}\lambda\mu$ times the probability of ruin at the first ladder time given there is a first ladder epoch. The expected number of ladder times is $(c - \lambda\mu)^{-1}\lambda\mu$. Intuitively, for u large, ruin will occur if one of the ladder heights is larger than u. ■

Proof Let $B(x)$ denote the distribution function of the first ladder height L_1, i.e.

$$B(x) = \frac{1}{\mu}\int_0^x (1 - G(y))\, dy\,.$$

Choose $\varepsilon > 0$ such that $\lambda\mu(1+\varepsilon) < c$. By Lemma F.6 there exists a D such that

$$\frac{1 - B^{*n}(x)}{1 - B(x)} \le D(1+\varepsilon)^n .$$

From the Pollaczek–Khintchine formula (5.10) we obtain

$$\begin{aligned}\frac{\psi(u)}{1 - B(u)} &= \Bigl(1 - \frac{\lambda\mu}{c}\Bigr)\sum_{n=1}^{\infty}\Bigl(\frac{\lambda\mu}{c}\Bigr)^n \frac{1 - B^{*n}(u)}{1 - B(u)} \\ &\le D\Bigl(1 - \frac{\lambda\mu}{c}\Bigr)\sum_{n=1}^{\infty}\Bigl(\frac{\lambda\mu}{c}\Bigr)^n (1+\varepsilon)^n < \infty .\end{aligned}$$

Thus we can interchange sum and limit. Recall from Lemma F.7 that

$$\lim_{u\to\infty}\frac{1 - B^{*n}(u)}{1 - B(u)} = n .$$

Thus

$$\begin{aligned}\lim_{u\to\infty}\frac{\psi(u)}{1 - B(u)} &= \Bigl(1 - \frac{\lambda\mu}{c}\Bigr)\sum_{n=1}^{\infty} n\Bigl(\frac{\lambda\mu}{c}\Bigr)^n = \Bigl(1 - \frac{\lambda\mu}{c}\Bigr)\sum_{n=1}^{\infty}\sum_{m=1}^{n}\Bigl(\frac{\lambda\mu}{c}\Bigr)^n \\ &= \Bigl(1 - \frac{\lambda\mu}{c}\Bigr)\sum_{m=1}^{\infty}\sum_{n=m}^{\infty}\Bigl(\frac{\lambda\mu}{c}\Bigr)^n = \sum_{m=1}^{\infty}\Bigl(\frac{\lambda\mu}{c}\Bigr)^m = \frac{\lambda\mu/c}{1 - \lambda\mu/c} \\ &= \frac{\lambda\mu}{c - \lambda\mu} .\end{aligned}$$

□

Example 5.28 Let $G \sim \text{Pa}(\alpha, \beta)$. Then (see Example F.5)

$$\lim_{x\to\infty}\frac{1 - B(zx)}{1 - B(x)} = z\lim_{x\to\infty}\frac{1 - G(zx)}{1 - G(x)} = z^{-(\alpha-1)} .$$

By Lemma F.4, B is a subexponential distribution. Note that we assume that $\mu < \infty$ and thus $\alpha > 1$. Because

$$\int_x^{\infty}\Bigl(\frac{\beta}{\beta+y}\Bigr)^{\alpha}\,\mathrm{d}y = \frac{\beta}{\alpha-1}\Bigl(\frac{\beta}{\beta+x}\Bigr)^{\alpha-1}$$

we obtain

$$\psi(u) \approx \frac{\lambda\beta/(\alpha-1)}{c - \lambda\beta/(\alpha-1)}\Bigl(\frac{\beta}{\beta+u}\Bigr)^{\alpha-1} = \frac{\lambda\beta}{c(\alpha-1) - \lambda\beta}\Bigl(\frac{\beta}{\beta+u}\Bigr)^{\alpha-1} .$$

Table 5.5 Approximations for subexponential claim sizes

u	$\psi(u)$	App	Er
1	0.364	8.79×10^{-3}	–97.588
2	0.150	1.52×10^{-4}	–99.898
3	6.18×10^{-2}	8.58×10^{-6}	–99.986
4	2.55×10^{-2}	9.22×10^{-7}	–99.996
5	1.05×10^{-2}	1.49×10^{-7}	–99.999
10	1.24×10^{-4}	3.47×10^{-10}	–100
20	1.75×10^{-8}	5.40×10^{-13}	–99.997
30	2.50×10^{-12}	1.10×10^{-14}	–99.56
40	1.60×10^{-15}	6.71×10^{-16}	–58.17
50	1.21×10^{-16}	7.56×10^{-17}	–37.69

Choose now $c = 1$, $\lambda = 9$, $\alpha = 11$ and $\beta = 1$. Table 5.5 gives the exact value ($\psi(u)$), the approximation (App) and the relative error in percent (Er). Consider for instance $u = 20$. The ruin probability is so small that it is no longer interesting for practical purposes. However, the approximation still underestimates the true value by almost 100%. That means we are still not far out enough in the tail. This is the problem with using the approximation. It should, however, be remarked that for small values of α the approximation is much better. Values $\alpha \in (1, 2)$ are also more interesting from a practical point of view. ■

Remark 5.29 The conditions of the theorem are also fulfilled for $\mathrm{LN}(\mu, \sigma^2)$, for $\mathrm{LG}(\gamma, \alpha)$ and for $\mathrm{Wei}(\alpha, c)$ ($\alpha < 1$) distributed claims, see [54, 92]. ■

5.12 The Time to Ruin

Consider the function

$$f_\alpha(u) = \mathbb{E}[\mathrm{e}^{-\alpha\tau} \mathbb{1}_{\{\tau<\infty\}} \mid C_0 = u]\,.$$

The function is defined at least for $\alpha \ge 0$. We will first find a differential equation for $f_\alpha(u)$.

Lemma 5.30 *The function $f_\alpha(u)$ is absolutely continuous and satisfies the equation*

$$cf_\alpha'(u) + \lambda\Bigl[\int_0^u f_\alpha(u-y)\,\mathrm{d}G(y) + 1 - G(u) - f_\alpha(u)\Bigr] - \alpha f_\alpha(u) = 0\,. \qquad (5.14)$$

Proof For h small we get

$$f_\alpha(u) = \int_0^h \Bigl[\int_0^{u+ct} e^{-\alpha t} f_\alpha(u+ct-y)\, dG(y) + e^{-\alpha t}(1-G(u+ct))\Bigr]\lambda e^{-\lambda t}\, dt \\ + e^{-\lambda h} e^{-\alpha h} f_\alpha(u+ch)\;.$$

We see that $f_\alpha(u)$ is right continuous. Reordering of the terms and dividing by h yields

$$0 = c\frac{f_\alpha(u+ch)-f_\alpha(u)}{ch} - \frac{1-e^{-(\lambda+\alpha)h}}{h} f_\alpha(u+ch) \\ + \frac{1}{h}\int_0^h \Bigl[\int_0^{u+ct} f_\alpha(u+ct-y)\, dG(y) + (1-G(u+ct))\Bigr]\lambda e^{-(\lambda+\alpha)t}\, dt\;.$$

Letting $h \to 0$ shows that $f_\alpha(u)$ is differentiable and that (5.14) for the derivative from the right holds. Replacing u by $u - ch$ gives the derivative from the left. □

This differential equation is hard to solve. Let us take the Laplace transform with respect to the initial capital. Let $\hat{f}_\alpha(s) = \int_0^\infty e^{-su} f_\alpha(u)\, du$. For the moment we assume $s > 0$. Note that (see Sect. 5.9)

$$\int_0^\infty f_\alpha'(u) e^{-su}\, du = s\hat{f}_\alpha(s) - f_\alpha(0)\;,$$

$$\int_0^\infty \int_0^u f_\alpha(u-y)\, dG(y) e^{-su}\, du = \hat{f}_\alpha(s) M_Y(-s)$$

and

$$\int_0^\infty \int_u^\infty dG(y) e^{-su}\, du = \int_0^\infty \int_0^y e^{-su}\, du\, dG(y) = \frac{1-M_Y(-s)}{s}\;.$$

Multiplying (5.14) by e^{-su} and integrating yields

$$c(s\hat{f}_\alpha(s) - f_\alpha(0)) + \lambda\Bigl[\hat{f}_\alpha(s) M_Y(-s) + \frac{1-M_Y(-s)}{s} - \hat{f}_\alpha(s)\Bigr] - \alpha\hat{f}_\alpha(s) = 0\;.$$

Solving for $\hat{f}_\alpha(s)$ yields

$$\hat{f}_\alpha(s) = \frac{cf_\alpha(0) - \lambda s^{-1}(1-M_Y(-s))}{cs - \lambda(1-M_Y(-s)) - \alpha}\;. \tag{5.15}$$

We know that $\hat{f}_\alpha(s)$ exists if $\alpha > 0$ and $s > 0$ and is positive. The denominator

$$cs - \lambda(1-M_Y(-s)) - \alpha$$

is convex, has value $-\alpha < 0$ at 0 and converges to ∞ as $s \to \infty$. Thus there exists a strictly positive root $s(\alpha)$ of the denominator. Because $\hat{f}_\alpha(s)$ exists for $s = s(\alpha)$ the numerator must have a root at $s(\alpha)$, too. Thus

$$cf_\alpha(0) = \lambda s(\alpha)^{-1}(1 - M_Y(-s(\alpha))) .$$

The function $s(\alpha)$ is differentiable by the implicit function theorem

$$s'(\alpha)(c - \lambda M_Y'(-s(\alpha))) - 1 = 0 . \tag{5.16}$$

Because $s(0) = 0$ we obtain that $\lim_{\alpha\to 0} s(\alpha) = 0$.

Example 5.31 Let the claims be Exp(β) distributed. We have to solve

$$cs - \frac{\lambda s}{\beta + s} - \alpha = 0 ,$$

which admits the two solutions

$$s_\pm(\alpha) = \frac{-(\beta c - \lambda - \alpha) \pm \sqrt{(\beta c - \lambda - \alpha)^2 + 4\alpha\beta c}}{2c} ,$$

where $s_-(\alpha) < 0 \le s_+(\alpha)$. Thus

$$\hat{f}_\alpha(s) = \frac{\lambda/(\beta + s_+(\alpha)) - \lambda/(\beta + s)}{cs - \lambda s/(\beta + s) - \alpha} = \frac{\lambda}{c(\beta + s_+(\alpha))(s - s_-(\alpha))} .$$

It follows that

$$f_\alpha(u) = \frac{\lambda}{c(\beta + s_+(\alpha))} e^{s_-(\alpha) u} .$$

■

Noting that

$$\mathbb{E}[\tau \mathbb{I}_{\{\tau<\infty\}}] = \lim_{\alpha\downarrow 0} \mathbb{E}[\tau e^{-\alpha\tau} \mathbb{I}_{\{\tau<\infty\}}] = \lim_{\alpha\downarrow 0} -\frac{d}{d\alpha} \mathbb{E}[e^{-\alpha\tau} \mathbb{I}_{\{\tau<\infty\}}]$$

we see that

$$\begin{aligned}\int_0^\infty \mathbb{E}[\tau \mathbb{I}_{\{\tau<\infty\}} \mid C_0 = u] e^{-su} \, du &= \lim_{\alpha\downarrow 0} -\frac{d}{d\alpha} \int_0^\infty f_\alpha(u) e^{-su} \, du \\ &= \lim_{\alpha\downarrow 0} -\frac{d}{d\alpha} \hat{f}_\alpha(s) .\end{aligned}$$

We can find the following "explicit" formula.

Lemma 5.32 *Assume $\mu_2 < \infty$. Then*

$$\mathbb{E}[\tau \mathbb{I}_{\{\tau<\infty\}}] = \frac{1}{c-\lambda\mu}\Big[\frac{\lambda\mu_2}{2(c-\lambda\mu)}\delta(u) - \int_0^u \psi(u-y)\delta(y)\,\mathrm{d}y\Big]. \tag{5.17}$$

Proof We get

$$\begin{aligned}-\frac{\mathrm{d}}{\mathrm{d}\alpha}\hat{f}_\alpha(s) &= \frac{\lambda s'(\alpha)}{cs-\lambda(1-M_Y(-s))-\alpha}\frac{1-M_Y(-s(\alpha))-M_Y'(-s(\alpha))s(\alpha)}{s(\alpha)^2}\\ &\quad - \frac{\lambda((1-M_Y(-s(\alpha)))s(\alpha)^{-1}-(1-M_Y(-s))s^{-1})}{(cs-\lambda(1-M_Y(-s))-\alpha)^2}. \end{aligned}\tag{5.18}$$

It follows from (5.16) that

$$s'(0) = \frac{1}{c-\lambda\mu}$$

and we have already seen that

$$\lim_{\alpha\downarrow 0}\frac{1-M_Y(-s(\alpha))}{s(\alpha)} = \lim_{s\downarrow 0}\frac{1-M_Y(-s)}{s} = \lim_{s\downarrow 0}M_Y'(-s) = \mu\,.$$

Moreover,

$$\begin{aligned}\lim_{\alpha\downarrow 0}\frac{(1-M_Y(-s(\alpha)))-M_Y'(-s(\alpha))s(\alpha)}{s(\alpha)^2} &= \lim_{s\downarrow 0}\frac{(1-M_Y(-s))-M_Y'(-s)s}{s^2}\\ &= \lim_{s\downarrow 0}\frac{sM_Y''(-s)}{2s} = \frac{\mu_2}{2}\,.\end{aligned}$$

Letting α tend to 0 in (5.18) yields

$$\begin{aligned}&\int_0^\infty \mathbb{E}[\tau\mathbb{I}_{\{\tau<\infty\}} \mid C_0 = u]\,\mathrm{e}^{-su}\,\mathrm{d}u\\ &\quad= \frac{\lambda\mu_2}{2(c-\lambda\mu)(cs-\lambda(1-M_Y(-s)))} - \frac{\lambda\mu-\lambda s^{-1}(1-M_Y(-s))}{(cs-\lambda(1-M_Y(-s)))^2}\\ &\quad= \frac{\lambda\mu_2}{2(c-\lambda\mu)^2}\frac{c-\lambda\mu}{cs-\lambda(1-M_Y(-s))}\\ &\qquad\quad - \frac{1}{c-\lambda\mu}\frac{c-\lambda\mu}{cs-\lambda(1-M_Y(-s))}\Big(\frac{1}{s}-\frac{c-\lambda\mu}{cs-\lambda(1-M_Y(-s))}\Big)\\ &\quad= \frac{1}{c-\lambda\mu}\Big[\frac{\lambda\mu_2}{2(c-\lambda\mu)}\hat{\delta}(s) - \hat{\delta}(s)\hat{\psi}(s)\Big].\end{aligned}$$

This is the Laplace transform of the assertion. □

Corollary 5.33 *Let $t > 0$. Then*

$$\mathbb{P}[t < \tau < \infty] < \frac{\lambda\mu_2}{2(c - \lambda\mu)^2 t}\,.$$

Proof Because $\delta(u)$ and $\psi(u)$ take values in $[0, 1]$ it is clear from (5.17) that

$$\mathbb{E}[\tau \mathbb{1}_{\{\tau<\infty\}}] < \frac{\lambda\mu_2}{2(c - \lambda\mu)^2}\,.$$

By Markov's inequality

$$\mathbb{P}[t < \tau < \infty] = \mathbb{P}[\tau \mathbb{1}_{\{\tau<\infty\}} > t] < \frac{1}{t}\mathbb{E}[\tau \mathbb{1}_{\{\tau<\infty\}}] < \frac{\lambda\mu_2}{2(c - \lambda\mu)^2 t}\,.$$

□

From (5.17) it is possible to get an explicit expression for $u = 0$

$$\mathbb{E}[\tau \mathbb{1}_{\{\tau<\infty\}} \mid C_0 = 0] = \frac{\lambda\mu_2}{2(c - \lambda\mu)^2}\Bigl(1 - \frac{\lambda\mu}{c}\Bigr) = \frac{\lambda\mu_2}{2c(c - \lambda\mu)}$$

and

$$\mathbb{E}[\tau \mid \tau < \infty, C_0 = 0] = \frac{\mu_2}{2\mu(c - \lambda\mu)}\,.$$

Example 5.34 For Exp(β) distributed claims we get for $R = \beta - \lambda/c$

$$\begin{aligned}
&\mathbb{E}[\tau \mathbb{1}_{\{\tau<\infty\}}]\\
&= \frac{\beta}{c\beta - \lambda}\Bigl[\frac{2\lambda}{2\beta(c\beta - \lambda)}\Bigl(1 - \frac{\lambda}{\beta c}\mathrm{e}^{-Ru}\Bigr) - \int_0^u \frac{\lambda}{\beta c}\mathrm{e}^{-R(u-y)}\Bigl(1 - \frac{\lambda}{\beta c}\mathrm{e}^{-Ry}\Bigr)\,\mathrm{d}y\Bigr]\\
&= \frac{\beta}{c\beta - \lambda}\Bigl[\frac{\lambda}{\beta(c\beta - \lambda)}\Bigl(1 - \frac{\lambda}{\beta c}\mathrm{e}^{-Ru}\Bigr) - \frac{\lambda}{\beta c}\mathrm{e}^{-Ru}\Bigl(\frac{c}{\beta c - \lambda}(\mathrm{e}^{Ru} - 1) - \frac{\lambda}{\beta c}u\Bigr)\Bigr]\\
&= \frac{\lambda}{\beta c^2(\beta c - \lambda)}\mathrm{e}^{-Ru}(\lambda u + c) = \frac{1}{c(\beta c - \lambda)}\psi(u)(\lambda u + c)\,.
\end{aligned}$$

The conditional expectation of the time of ruin is linear in u

$$\mathbb{E}[\tau \mid \tau < \infty] = \frac{\lambda u + c}{c(\beta c - \lambda)}\,.$$

■

5.13 Seal's Formulae

We consider now the probability of ruin within finite time $\psi(u, t)$. However, first we want to find the conditional finite time ruin probability given C_t for some t fixed.

Lemma 5.35 *Let t be fixed, $u = 0$ and $0 < y \le ct$. Then*

$$\mathbb{P}[C_s \ge 0, 0 \le s \le t \mid C_t = y] = \frac{y}{ct}\,.$$

Proof Consider

$$C_t - C_{(t-s)-} = cs - \sum_{i=N_{(t-s)-}+1}^{N_t} Y_i\,.$$

This is also a Cramér–Lundberg model. Thus

$$\begin{aligned}\mathbb{P}[C_s \ge 0, 0 \le s \le t \mid C_t = y] &= \mathbb{P}[C_t - C_{t-s} \ge 0, 0 \le s \le t \mid C_t - C_0 = y] \\ &= \mathbb{P}[C_s \le C_t, 0 \le s \le t \mid C_t = y]\,.\end{aligned}$$

Let $S_s = cs - C_s$ denote the aggregate claims up to time s. Denote by σ the permutations of $\{1, 2, \dots, n\}$. Then

$$\begin{aligned}&\mathbb{E}[S_s \mid S_t = y, N_t = n, N_s = k] \\ &\quad = \frac{1}{n!}\mathbb{E}\Bigl[\sum_\sigma \sum_{i=1}^k Y_{\sigma(i)} \Bigm| S_t = y, N_t = n, N_s = k\Bigr] \\ &\quad = \frac{k(n-1)!}{n!}\mathbb{E}\Bigl[\sum_{i=1}^n Y_i \Bigm| S_t = y, N_t = n, N_s = k\Bigr] = \frac{ky}{n}\,.\end{aligned}$$

Because, given $\{N_t = n\}$, the claim times are uniformly distributed in $[0, t]$ (Proposition C.3) we obtain

$$\begin{aligned}\mathbb{E}[S_s \mid S_t = y, N_t = n] &= \sum_{k=0}^n \binom{n}{k}\Bigl(\frac{s}{t}\Bigr)^k \Bigl(\frac{t-s}{t}\Bigr)^{n-k} \frac{ky}{n} \\ &= \sum_{k=1}^n \binom{n-1}{k-1}\Bigl(\frac{s}{t}\Bigr)^k \Bigl(\frac{t-s}{t}\Bigr)^{n-k} y = \frac{sy}{t}\,.\end{aligned}$$

This is independent of n. Thus

$$\mathbb{E}[S_s \mid S_t = y] = \frac{sy}{t}$$

and

$$\mathbb{E}[C_s \mid C_t = y] = \mathbb{E}[cs - S_s \mid S_t = ct - y] = cs - \frac{s(ct - y)}{t} = \frac{sy}{t}\,.$$

Then for $v \le s < t$

$$\begin{aligned}\mathbb{E}\Bigl[\frac{y - C_s}{t - s} \Bigm| C_v, C_t = y\Bigr] &= \frac{y - C_v - \mathbb{E}[(C_s - C_v) \mid C_v, C_t = y]}{t - s} \\ &= \frac{y - C_v - (s - v)(y - C_v)/(t - v)}{t - s} = \frac{y - C_v}{t - v}\end{aligned}$$

and the process

$$M_s = \frac{y - C_s}{t - s} \qquad (0 \le s < t)$$

is a conditional martingale. Note that $\lim_{s\uparrow t} M_s = c$. Let $T = \inf\{s \ge 0 : C_s = y\}$. Because

$$0 \le M_s \mathbb{I}_{\{T>s\}} \le c$$

on $\{C_t = y\}$ we have

$$\begin{aligned}\lim_{s\uparrow t} \mathbb{E}[M_s \mathbb{I}_{\{T>s\}} \mid C_t = y] &= \mathbb{E}\Bigl[\lim_{s\uparrow t} M_s \mathbb{I}_{\{T>s\}} \Bigm| C_t = y\Bigr] \\ &= c\mathbb{P}[T = t \mid C_t = y]\,.\end{aligned}$$

Note that by the stopping theorem $\{M_{T\wedge t}\}$ is a bounded martingale. Thus because $M_T = 0$ on $\{T < t\}$

$$\begin{aligned}\frac{y}{t} &= M_0 = \lim_{s\uparrow t} \mathbb{E}[M_{T\wedge s} \mid C_t = y] = \lim_{s\uparrow t} \mathbb{E}[M_s \mathbb{I}_{\{T>s\}} \mid C_t = y] \\ &= c\mathbb{P}[T = t \mid C_t = y]\,.\end{aligned}$$

Note that $\mathbb{P}[T = t \mid C_t = y] = \mathbb{P}[C_s \le C_t, 0 \le s \le t \mid C_t = y]$. □

Conditioning on C_t is in fact a conditioning on the aggregate claim size S_t. Denote by $F(x; t)$ the distribution function of S_t. For the integration we use $\mathrm{d}F(\cdot; t)$ for an integration with respect to the measure $F(\cdot; t)$. Moreover, let $\delta(u, t) = 1 - \psi(u, t) = \mathbb{P}[\tau > t]$.

Theorem 5.36 *For initial capital 0 we have*

$$\delta(0, t) = \frac{1}{ct}\mathbb{E}[C_t \vee 0] = \frac{1}{ct}\int_0^{ct} F(y; t)\,\mathrm{d}y\,.$$

Proof By Lemma 5.35 it follows readily that

$$\delta(0,t) = \mathbb{E}[\mathbb{P}[\tau > t \mid C_t]] = \mathbb{E}\Big[\frac{C_t \vee 0}{ct}\Big].$$

Moreover,

$$\begin{aligned}\mathbb{E}[C_t \vee 0] &= \mathbb{E}[(ct - S_t) \vee 0] = \int_0^{ct} (ct - z)\, \mathrm{d}F(z;t) \\ &= \int_0^{ct}\int_z^{ct} \mathrm{d}y\, \mathrm{d}F(z;t) = \int_0^{ct}\int_0^{y} \mathrm{d}F(z;t)\, \mathrm{d}y = \int_0^{ct} F(y;t)\, \mathrm{d}y\,.\end{aligned}$$

□

Now let the initial capital be arbitrary.

Theorem 5.37 *With the notation used above we have for* $u > 0$

$$\delta(u,t) = F(u+ct;t) - \int_u^{u+ct} \delta\Big(0, t - \frac{v-u}{c}\Big) F\Big(\mathrm{d}v; \frac{v-u}{c}\Big)\,.$$

Proof Note that

$$\delta(u,t) = \mathbb{P}[C_t > 0] - \mathbb{P}[\exists 0 \le s < t : C_s = 0, C_v > 0 \text{ for } s < v \le t]$$

and $\mathbb{P}[C_t > 0] = F(u+ct;t)$. Let $T = \sup\{0 \le s \le t : C_s = 0\}$ and set $T = \infty$ if $C_s > 0$ for all $s \in [0,t]$. Then, noting that $\mathbb{P}[C_s > 0 : 0 < s \le t \mid C_0 = 0] = \mathbb{P}[C_s \ge 0 : 0 < s \le t \mid C_0 = 0]$,

$$\begin{aligned}\mathbb{P}[T \in [s, s+\mathrm{d}s)] &= \mathbb{P}[C_s \in (-c\,\mathrm{d}s, 0], C_v > 0 \text{ for } v \in [s+\mathrm{d}s, t]] \\ &= (F(u + c(s+\mathrm{d}s); s) - F(u+cs;s))\delta(0, t-s-\mathrm{d}s) \\ &= \delta(0,t-s)c\, \mathrm{d}F(u+cs;s)\,.\end{aligned}$$

Thus

$$\begin{aligned}\delta(u,t) &= F(u+ct;t) - \int_0^t \delta(0,t-s)c\, \mathrm{d}F(u+cs;s) \\ &= F(u+ct;t) - \int_u^{u+ct} \delta\Big(0, t - \frac{v-u}{c}\Big)\, \mathrm{d}F\Big(v, \frac{v-u}{c}\Big)\,.\end{aligned}$$

□

5.14 Finite Time Lundberg Inequalities

In this section we derive an upper bound for probabilities of the form

$$\mathbb{P}[\underline{y}u < \tau \le \bar{y}u]$$

for $0 \le \underline{y} < \bar{y} < \infty$. Assume that the Lundberg exponent R exists. We will use the martingale (5.4) and the stopping theorem. For $r \ge 0$ such that $M_Y(r) < \infty$,

$$\begin{aligned} e^{-ru} &= \mathbb{E}[e^{-rC_{\tau\wedge\bar{y}u}-\theta(r)(\tau\wedge\bar{y}u)}] > \mathbb{E}[e^{-rC_{\tau\wedge\bar{y}u}-\theta(r)(\tau\wedge\bar{y}u)}; \underline{y}u < \tau \le \bar{y}u] \\ &= \mathbb{E}[e^{-rC_\tau-\theta(r)\tau} \mid \underline{y}u < \tau \le \bar{y}u]\mathbb{P}[\underline{y}u < \tau \le \bar{y}u] \\ &> \mathbb{E}[e^{-\theta(r)\tau} \mid \underline{y}u < \tau \le \bar{y}u]\mathbb{P}[\underline{y}u < \tau \le \bar{y}u] \\ &> e^{-\max\{\theta(r)\underline{y}u,\theta(r)\bar{y}u\}}\mathbb{P}[\underline{y}u < \tau \le \bar{y}u]\,. \end{aligned}$$

Thus

$$\mathbb{P}[\underline{y}u < \tau \le \bar{y}u \mid C_0 = u] < e^{-(r-\max\{\theta(r)\underline{y},\theta(r)\bar{y}\})u}\,.$$

Choosing the exponent as small as possible we obtain

$$\mathbb{P}[\underline{y}u < \tau \le \bar{y}u \mid C_0 = u] < e^{-R(\underline{y},\bar{y})u}\,, \tag{5.19}$$

where

$$R(\underline{y},\bar{y}) = \sup_{r\in\mathbb{R}} \min\{r - \theta(r)\underline{y}, r - \theta(r)\bar{y}\}\,.$$

The supremum over $r \in \mathbb{R}$ makes sense because $\theta(r) > 0$ for $r < 0$ and $\theta(r) = \infty$ if $M_Y(r) = \infty$. Since $\theta(R) = 0$ it follows that $R(\underline{y},\bar{y}) \ge R$. Hence (5.19) could be useful.

For further investigation of $R(\underline{y},\bar{y})$ we consider the function $f_y(r) = r - y\theta(r)$. Let $r_\infty = \sup\{r \ge 0 : M_Y(r) < \infty\}$. Then $f_y(r)$ exists in the interval $(-\infty, r_\infty)$ and $f_y(r) = -\infty$ for $r \in (r_\infty, \infty)$. If $M_Y(r_\infty) < \infty$ then $|f_y(r_\infty)| < \infty$. Since $f_y''(r) = -yM_Y''(r) < 0$ the function $f_y(r)$ is concave. Thus there exists a unique r_y such that $f_y(r_y) = \sup\{f_y(r) : r \in \mathbb{R}\}$. Because $f_y(0) = 0$ and $f_y'(0) = 1 - y\theta'(0) = 1 + y(c - \lambda\mu) > 0$ we conclude that $r_y > 0$. r_y is either the solution to the equation

$$1 + y(c - \lambda M_Y'(r_y)) = 0$$

or $r_y = r_\infty$. We assume now that $R < r_\infty$ and let

$$y_0 = \left(\lambda M_Y'(R) - c\right)^{-1}\,.$$

It follows that $r_{y_0} = R$. We call y_0 the *critical value*.

Because $\frac{\mathrm{d}}{\mathrm{d}y} f_y(r) = -\theta(r)$ it follows readily that

$$\frac{\mathrm{d}}{\mathrm{d}y} f_y(r) \gtreqless 0 \iff r \lesseqgtr R\,.$$

We get

$$f_{\underline{y}}(r) \lesseqgtr f_{\bar{y}}(r) \text{ as } r \lesseqgtr R\,.$$

Because $M_Y'(r)$ is an increasing function it follows that

$$y \gtreqless y_0 \iff r_y \lesseqgtr R\,.$$

Thus

$$R(\underline{y}, \bar{y}) = \begin{cases} R & \text{if } \ \underline{y} \le y_0 \le \bar{y}, \\ f_{\bar{y}}(r_{\bar{y}}) & \text{if } \ \bar{y} < y_0, \\ f_{\underline{y}}(r_{\underline{y}}) & \text{if } \ \underline{y} > y_0. \end{cases}$$

If $y_0 \in [\underline{y}, \bar{y}]$ then we again get Lundberg's inequality. If $\bar{y} < y_0$ then $R(\underline{y}, \bar{y})$ does not depend on $\underline{y}$. By choosing $\underline{y}$ as small as possible we obtain

$$\mathbb{P}[0 < \tau \le yu \mid C_0 = u] < \mathrm{e}^{-R(0,y)u} \tag{5.20}$$

for $y < y_0$. Note that $R(0, y) > R$. Analogously

$$\mathbb{P}[yu < \tau < \infty \mid C_0 = u] < \mathrm{e}^{-R(y,\infty)u} \tag{5.21}$$

for $y > y_0$. The strict inequality follows from

$$\mathrm{e}^{-ru} \ge \mathbb{E}[\mathrm{e}^{-rC_\tau - \theta(r)\tau}; yu < \tau < \infty] > \mathrm{e}^{-\theta(r)yu}\,\mathbb{P}[yu < \tau < \infty]$$

if $0 < r < R$, compare also (5.5). Again $R(y, \infty) > R$.

We see that $\mathbb{P}[\tau \notin ((y_0 - \varepsilon)u, (y_0 + \varepsilon)u)]$ tends to 0 faster than $\mathbb{P}[\tau \in ((y_0 - \varepsilon)u, (y_0 + \varepsilon)u)]$. Intuitively ruin should therefore occur close to $y_0 u$.

Theorem 5.38 *Assume that* $R < r_\infty$. *Then*

$$\frac{\tau}{u} \longrightarrow y_0$$

in probability on the set $\{\tau < \infty\}$ *as* $u \to \infty$.

Proof Choose $\varepsilon > 0$. By the Cramér–Lundberg approximation $\mathbb{P}[\tau < \infty \mid C_0 = u]\exp\{Ru\} \to C$ for some $C > 0$. Then

$$\begin{aligned}
&\mathbb{P}\Big[\Big|\frac{\tau}{u} - y_0\Big| > \varepsilon \,\Big|\, \tau < \infty, C_0 = u\Big] \\
&\quad = \frac{\mathbb{P}[\tau < (y_0 - \varepsilon)u \mid C_0 = u] + \mathbb{P}[(y_0 + \varepsilon)u < \tau < \infty \mid C_0 = u]}{\mathbb{P}[\tau < \infty \mid C_0 = u]} \\
&\quad \le \frac{\exp\{-R(0, y_0 - \varepsilon)u\} + \exp\{-R(y_0 + \varepsilon, \infty)u\}}{\mathbb{P}[\tau < \infty \mid C_0 = u]} \\
&\quad = \frac{\exp\{-(R(0, y_0 - \varepsilon) - R)u\} + \exp\{-(R(y_0 + \varepsilon, \infty) - R)u\}}{\mathbb{P}[\tau < \infty \mid C_0 = u]e^{Ru}} \longrightarrow 0
\end{aligned}$$

as $u \to \infty$. □

5.15 Capital Injections

A disadvantage of measuring the risk via ruin probabilities is that they neither take into account the time to ruin nor the deficit at ruin. A simple idea is to consider a Gerber–Shiu function $\mathbb{E}[e^{-\delta\tau}|C_\tau|]$. Then both the deficit and the time to ruin are taken into account. Here we go a step further and do not allow ruin. If ruin occurs, a capital injection has to be made and the process proceeds. We then measure the risk by the discounted value of the capital injections. Like for ruin probabilities, since the law remains the same over time, capital injections also have to be considered as a technical term. Further, the discounting has to be seen as a preference measure, injections tomorrow are preferred to injections today.

Let $L_t = \sup\{C_s^- : 0 \le s \le t\}$. Then L is the minimal process such that $C_t^L := C_t + L_t \ge 0$ for all t. The discounted value of the capital injections is

$$V(x) = \mathbb{E}\Big[\int_{0-}^\infty e^{-\delta t}\, dL_t\Big] = \mathbb{E}\Big[\int_0^\infty \delta e^{-\delta t} L_t\, dt\Big] .$$

It is clear that $V(x)$ is decreasing in x, $\lim_{x\to\infty} V(x) = 0$, and $V(x) = V(0) - x$ if $x < 0$.

Lemma 5.39 *It is not optimal to inject capital if the surplus is positive.*

Proof Suppose $u \ge 0$ and $z = L_0 > 0$. Assume first that $\delta > 0$. Consider an arbitrary strategy with $L_0 = z$ and the alternative strategy $\tilde{L}_t = L_t \mathbb{I}_{t<T_1}$. Then

$$V(u, L) - \tilde{V}(u, \tilde{L}) = \mathbb{E}\Big[\int_0^{T_1} \delta e^{-\delta t} L_t\, dt\Big] \ge z\mathbb{E}[1 - e^{-\delta T_1}] = \frac{z\delta}{\lambda + \delta} .$$

Maximising over all strategies yields $V(u) - \tilde{V}(u) \ge z\delta/(\lambda + \delta)$, and the strategy $\tilde{L}_t$ is preferred. Assume now that $\delta = 0$. Then $V(x) = \mathbb{E}[L_\infty]$. Consider an arbitrary strategy L with $L_0 = z$ and the alternative strategy $\tilde{L}_t = \mathbb{I}_{\tau \le t} L_t$, where τ is as before the time to ruin without injections. Then $V(x, L) - \tilde{V}(x, \tilde{L}) = V(x, L)\mathbb{P}[\tau = \infty]$.

Maximising over all strategies yields $V(u) - \tilde{V}(u) = V(u)\mathbb{P}[\tau = \infty]$. Because $\mathbb{P}[\tau = \infty] < 1$, we get the result. □

Lemma 5.40 *The value function $V(x)$ is strictly convex.*

Proof Let $x, y \geq 0$ be two initial capitals and L^x and L^y, respectively, be the corresponding capital injections process. Then for initial capital $z = \alpha x + (1-\alpha)y$ we have

$$C_t^z + \alpha L_t^x + (1-\alpha)L_t^y = \alpha(C_t^x + L_t^x) + (1-\alpha)(C_t^y + L_t^y) \geq 0\;.$$

This implies that $L_t^z \leq \alpha L_t^x + (1-\alpha)L_t^y$, and

$$\begin{aligned} V(z) &= \mathbb{E}\Bigl[\int_0^\infty \delta e^{-\delta t} L_t^z \,\mathrm{d}t\Bigr] \\ &\leq \alpha\mathbb{E}\Bigl[\int_0^\infty \delta e^{-\delta t} L_t^x \,\mathrm{d}t\Bigr] + (1-\alpha)\mathbb{E}\Bigl[\int_0^\infty \delta e^{-\delta t} L_t^y \,\mathrm{d}t\Bigr] \\ &= \alpha V(x) + (1-\alpha)V(y)\;. \end{aligned}$$

□

We will show below that $V(x)$ is continuously differentiable. Another property that follows readily is that $V'(x) > -1$ for $x > 0$. Indeed, if $V'(x) < -1$ for some x, this must hold on some interval. Then it would be optimal to inject capital immediately, which we have seen is not optimal. Thus $V'(x) \geq -1$ for all x. If $V'(x) = -1$ for some $x > 0$, then $V'(y) = -1$ on $[0, x]$ by the convexity of $V(x)$. Then for any $y < x$ an immediate capital injection of size $x - y$ has the same value as no injection, but this is not optimal either.

The approach for finding integro-differential equations we have used above gives the following.

Theorem 5.41 *$V(x)$ is the unique solution to the equation*

$$cV'(x) + \lambda\Bigl[\int_0^x V(x-y)\,\mathrm{d}G(y) + \int_x^\infty (V(0)+y-x)\,\mathrm{d}G(y)\Bigr] - (\lambda+\delta)V(x) = 0\;.$$

Remark 5.42 Since $V(x)$ is continuous at zero, it turns out that $V(x)$ is continuously differentiable even if $G(x)$ is not continuous. Note further that

$$\int_x^\infty (V(0)+y-x)\,\mathrm{d}G(y) = V(0)(1-G(x)) + \int_x^\infty (1-G(y))\,\mathrm{d}y\;.$$

This implies in particular that $V'(0) = c^{-1}(\delta V(0) - \lambda\mu)$. Since $V(x)$ is decreasing, this implies $V(0) \leq \delta^{-1}\lambda\mu$. ■

Let now V_0 be arbitrary and consider the solution to

$$cf'(x)+\lambda\Bigl[\int_0^x f(x-y)\,\mathrm{d}G(y)+\int_x^\infty (V_0+y-x)\,\mathrm{d}G(y)\Bigr]-(\lambda+\delta)f(x)=0 \quad (5.22)$$

with the property that $\lim_{x\to\infty} f(x)=0$. Consider further $\tilde{f}(x)=\mathbb{E}[(V_0-C_\tau)\mathrm{e}^{-\delta\tau}]$. Then $\tilde{f}(x)$ also satisfies (5.22). The function $g(x)=f(x)-\tilde{f}(x)$ then satisfies

$$cg'(x)+\lambda\int_0^x g(x-y)\,\mathrm{d}G(y)-(\lambda+\delta)g(x)=0\,. \quad (5.23)$$

If $g(0)>0$, then $g'(x)>0$ for all $x>0$, hence $g(\infty)>0$. This contradicts our assumption that $g(\infty)=0$. Analogously, $g(0)<0$ is not possible. Thus, $g(0)=0$. Taking the Laplace transform of (5.23), see also Sect. 5.9, yields

$$cs\hat{g}(s)+(\lambda M_Y(-s)-\lambda-\delta)\hat{g}(s)=0\,.$$

Since $cs+\lambda M_Y(-s)-\lambda-\delta\neq 0$, we conclude that $\hat{g}(s)=0$. This implies that $f(x)=\tilde{f}(x)$. In particular, we find

$$f(0)-V(0)=(V_0-V(0))\mathbb{E}[\mathrm{e}^{-\delta\tau}]\,.$$

In particular, $V_0 \gtreqqless V(0)$ if and only if $f(0)\lesseqqgtr V_0$. Moreover, starting with an arbitrary value, we can determine $V(0)$ in terms of $\mathbb{E}[\mathrm{e}^{-\delta\tau}]$. We considered the latter quantity in Sect. 5.12.

Example 5.43 Consider exponentially distributed claim sizes $G(x)=1-\mathrm{e}^{-\alpha x}$. Then, using the lack of memory property, we need to solve

$$cf'(x)+\lambda\mathrm{e}^{-\alpha x}\Bigl[\int_0^x f(y)\alpha\mathrm{e}^{\alpha y}\,\mathrm{d}y+V_0+\alpha^{-1}\Bigr]-(\lambda+\delta)f(x)=0\,.$$

Taking the derivative gives

$$cf''(x)+\alpha[cf'(x)-(\lambda+\delta)f(x)]+\lambda\alpha f(x)-(\lambda+\delta)f'(x)=0\,,$$

where we replace the term with the integral with the expression of the equation to solve. Thus, $f(x)=A\mathrm{e}^{\rho x}+B\mathrm{e}^{-Rx}$, where

$$\rho=\frac{\sqrt{(\alpha c-\lambda-\delta)^2+4\alpha\delta c}+\lambda+\delta-\alpha c}{2c}\,,$$
$$R=\frac{\sqrt{(\alpha c-\lambda-\delta)^2+4\alpha\delta c}+\alpha c-\lambda-\delta}{2c}\,.$$

In order that the solution converges as $x \to \infty$, we must have $A = 0$. Plugging Be^{-Rx} into the original equation we obtain

$$-Be^{-Rx}[cR + \lambda + \delta] + \frac{\alpha\lambda B}{\alpha - R}(e^{-Rx} - e^{-\alpha x}) + \lambda(V_0 - \alpha^{-1})e^{-\alpha x} = 0 .$$

The terms with e^{-Rx} disappear, giving the condition $V_0 - \alpha^{-1} - \alpha B/(\alpha - R) = 0$, or $B = (V_0 - \alpha^{-1})(\alpha - R)/\alpha$. The initial condition $V(0)$ we are interested in must satisfy $V(0) = (V(0) + \alpha^{-1})(\alpha - R)/\alpha$. This means $V(0) = (\alpha - R)/(\alpha R) = R^{-1} - \alpha^{-1}$. We find the value function $V(x) = (R^{-1} - \alpha^{-1})e^{-Rx}$. ■

Ruin probabilities as well as capital injections could be considered as measures for the risk. The supervisor could ask for a capital to keep the ruin probability under a certain level. That is,

$$\inf\{u : \mathbb{P}[\tau = \infty] \geq \alpha\} = \inf\{u : \mathbb{P}[\inf_t C_t \geq 0] \geq \alpha\} .$$

This is the Value-at-Risk applied to $\inf_t C_t$. Therefore, this risk measure has the same properties as the V@R. Another possibility would be to set the regulatory capital to $V(u)$. Because we are considering a stochastic process, we cannot directly use the properties of a risk measure detailed in Sect. 1.11. If $\tilde{C}_t \geq C_t$ for all t, then $\tilde{V}(u) \leq V(u)$. Thus the measure is monotone. If $\tilde{C}_t = x + C_t$, then because $V'(x) > -1$, we have $V(u+x) > V(u) - x$. Thus, even in the case $\delta = 0$, translation invariance is not satisfied. The case $\lambda = 0$ gives $V(x) = 0$. So not taking a risk does not require a regulatory capital. This is something like a normed measure.

Considering the process λC_t, we get $V_\alpha(\alpha u) = \alpha V(u)$. This gives that the measure is positive homogeneous. In the same way as we have proved Lemma 5.40, it follows that the value for the process $\alpha C_t + \beta \tilde{C}_t$ is smaller than $\alpha V(C_0) + \beta \tilde{V}(\tilde{C}_0)$, where $\tilde{V}(u)$ is the value corresponding to the process $\tilde{C}_t$. This holds for any $\alpha, \beta \geq 0$. Thus the measure is convex and also subadditive.

Bibliographical Remarks

Further literature on the classical risk model can be found, for example, in [11, 19, 28, 47, 65, 74, 78, 88, 110]. The Cramér–Lundberg model was introduced by Filip Lundberg [96]. The basic results, like Lundberg's inequality and the Cramér–Lundberg approximation, go back to Lundberg [97] and Cramér [36], see also [37, 74, 110]. Upper and lower bounds may be found, for example, in [110]. The differential equation for the ruin probability (5.2) and the Laplace transform of the ruin probability can, for instance, be found in [37]. The renewal theory proof of the Cramér–Lundberg approximation is due to Feller [59]. The martingale (5.4) was introduced by Gerber [62]. The latter paper also contains the martingale proof of Lundberg's inequality and the proof of (5.20). The proof of (5.21) goes back to Grandell [75]. Section 5.14 can also be found in [55]. Some similar results are obtained in [4, 5]. Segerdahl

[121] proved the asymptotic normality of the ruin time under the condition that ruin occurs.

Proposition 5.13 can be found in [84]. The results on reinsurance and ruin can be found in [33, 136]. The approach for proportional reinsurance as well as the example is from [83]. Proposition 5.13 can be found in [65, p. 130].

The term *severity of ruin* was introduced in [66], the joint distribution of the severity of ruin and the capital prior to ruin was first investigated in [49]. More results on the topic are obtained in [46, 110, 115]. A slight generalisation are the so-called Gerber–Shiu functions, see [68]. Further references can be found in [117]. A diffusion approximation, even though not mathematically correctly obtained, was already used by Hadwiger [82]. The modern approach goes back to Iglehart [86], see also [9, 73, 113]. The deVylder approximation was introduced in [44]. Formula (5.13) was found by Asmussen [9]. It is however stated incorrectly in the paper. For the correct formula, see [18] or [110]. The Beekman–Bowers approximation can be found in [20].

Theorem 5.26 was obtain by von Bahr [133] in the special case of Pareto distributed claim sizes and by Thorin and Wikstad [132] in the case of lognormally distributed claims. The theorem in the present form is essentially given in [54], see also [56].

Section 5.12 is taken from [112]. Parts can also be found in [18, 113]. Equation (5.15) was first obtained in [37]. The expected value of the ruin time for exponentially distributed claims can be found in [65, p. 138]. Seal's formulae were first obtained by Takács [125, p. 56]. Seal [119, 120] introduced them to risk theory. The present presentation follows [65].

Measuring the risk via capital injections was introduced in [51, 52]. A theory on risk functionals on stochastic processes can be found in [34], see also the references therein.

Many of the results of this chapter have been generalised to Lévy processes. One of the first references is [48]. Further literature may be found in the books [21, 94] and the references therein.

Chapter 6
The Renewal Risk Model

We now generalise some of the results of Chap. 5 to the case where the claims arrival process is a renewal process.

6.1 Definition of the Renewal Risk Model

The simplest generalisation of the classical risk model is to assume that the claim number process is a renewal process. Then the risk process is no longer Markovian, because the distribution of the time of the next claim depends on the past via the time of the last claim. We therefore cannot find a simple differential equation for $\psi(u)$ as in the classical case.

Let

$$C_t = u + ct - \sum_{i=1}^{N_t} Y_i ,$$

where

- $\{N_t\}$ is a renewal process with event times $0 = T_0, T_1, T_2, \dots$ with interarrival distribution function F, mean λ^{-1} and moment generating function $M_T(r)$. We denote by T a generic random variable with distribution function F. The first claim time T_1 has distribution function F^1. Unless stated otherwise, we consider the ordinary case where $F^1 = F$.
- The claim amounts $\{Y_i\}$ build an iid. sequence with distribution function G, moments $\mu_k = \mathbb{E}[Y_i^k]$ and moment generating function $M_Y(r)$. As in the classical model $\mu = \mu_1$.
- $\{N_t\}$ and $\{Y_i\}$ are independent.

This model is called the *renewal risk model* or, after the person who introduced the model, the *Sparre Andersen model*.

H. Schmidli, *Risk Theory*, Springer Actuarial,
DOI:10.1007/978-3-319-72005-0_6

As in the classical model we define the ruin time $\tau = \inf\{t > 0 : C_t < 0\}$ and the ruin probabilities $\psi(u,t) = \mathbb{P}[\tau \le t]$ and $\psi(u) = \mathbb{P}[\tau < \infty]$. Considering the random walk $\{C_{T_i}\}$ we can conclude that

$$\psi(u) = 1 \quad \Longleftrightarrow \quad \mathbb{E}[C_{T_2} - C_{T_1}] = c/\lambda - \mu \le 0\,.$$

In order to avoid ruin a.s. we assume the net profit condition $c > \lambda\mu$. Note that in contrast to the classical case

$$\mathbb{E}[C_t - u] \ne (c - \lambda\mu)t \quad \text{for most } t > 0$$

except if T has an exponential distribution. By the law of large numbers,

$$\lim_{t\to\infty} \frac{1}{t}(C_t - u) = c - \lambda\mu\,.$$

It follows again that $\psi(u) \to 0$ as $u \to \infty$ because $\inf\{C_t - u, t \ge 0\}$ is finite.

6.2 The Adjustment Coefficient

We were able to construct some martingales of the form $\{\exp\{-rC_t - \theta(r)t\}\}$ in the classical case. Because the risk process $\{C_t\}$ is no longer Markovian we cannot hope to get such a martingale again because Lemma 5.1 does not hold. However, the claim times are renewal times because $\{C_{T_1+t} - C_{T_1}\}$ is a renewal risk model independent of T_1 and Y_1.

Lemma 6.1 *Let $\{C_t\}$ be an (ordinary) renewal risk model. For any $r \in \mathbb{R}$ such that $\inf\{1/M_T(s) : s \ge 0, M_T(s) < \infty\} \le M_Y(r) < \infty$ let $\theta(r)$ be the unique solution to*

$$M_Y(r)M_T(-\theta(r) - cr) = 1\,. \tag{6.1}$$

Then the discrete time process

$$\{\exp\{-rC_{T_i} - \theta(r)T_i\}\}$$

is a martingale.

Remark 6.2 In the classical case with $F(t) = 1 - e^{-\lambda t}$ Eq. (6.1) is

$$M_Y(r)\frac{\lambda}{\lambda + \theta(r) + cr} = 1\,,$$

which is equivalent to Eq. (5.3) in the classical case. ∎

Proof Note that $1/M_Y(r) \le \sup\{M_T(s) : M_T(s) < \infty\}$. Because $M_T(r)$ is an increasing and continuous function defined for all $r \le 0$, and because $M_T(r) \to 0$ as $r \to -\infty$, there exists a unique solution $\theta(r)$ to (6.1). Then

$$\begin{aligned}
&\mathbb{E}[e^{-rC_{T_{i+1}}-\theta(r)T_{i+1}} \mid \mathcal{F}_{T_i}] \\
&\quad = \mathbb{E}[e^{-r(c(T_{i+1}-T_i)-Y_{i+1})-\theta(r)(T_{i+1}-T_i)} \mid \mathcal{F}_{T_i}]e^{-rC_{T_i}-\theta(r)T_i} \\
&\quad = \mathbb{E}[e^{rY_{i+1}}e^{-(cr+\theta(r))(T_{i+1}-T_i)}]e^{-rC_{T_i}-\theta(r)T_i} \\
&\quad = M_Y(r)M_T(-cr-\theta(r))e^{-rC_{T_i}-\theta(r)T_i} = e^{-rC_{T_i}-\theta(r)T_i}\,.
\end{aligned}$$

□

As in the classical case we are interested in the case $\theta(r) = 0$. In the classical case there existed, besides the trivial solution $r = 0$, at most a second solution to the equation $\theta(r) = 0$. This was the case because $\theta(r)$ was a convex function. Let us therefore compute the second derivative of $\theta(r)$. Define $m_Y(r) = \log M_Y(r)$ and $m_T(r) = \log M_T(r)$. Then $\theta(r)$ is the solution to the equation

$$m_Y(r) + m_T(-\theta(r) - cr) = 0\,.$$

By the implicit function theorem $\theta(r)$ is differentiable and

$$m_Y'(r) - (\theta'(r) + c)m_T'(-\theta(r) - cr) = 0\,. \tag{6.2}$$

Hence $\theta(r)$ is infinitely often differentiable and

$$m_Y''(r) - \theta''(r)m_T'(-\theta(r) - cr) + (\theta'(r) + c)^2 m_T''(-\theta(r) - cr) = 0\,.$$

Note that $M_T(r)$ is a strictly increasing function and so is $m_T(r)$. Thus $m_T'(-\theta(r) - cr) > 0$. By Lemma 1.13 both $m_Y''(r)$ and $m_T''(r)$ are positive. We can assume that Y and T are not both deterministic, otherwise $\psi(u) = 0$. Then at least one of the two functions $m_Y''(r)$ and $m_T''(r)$ is strictly positive. From (6.2) it follows that $\theta'(r) + c > 0$. Thus $\theta''(r) > 0$ and $\theta(r)$ is a strictly convex function.

Let $r = 0$, i.e. $\theta(0) = 0$. Then

$$\theta'(0) = \frac{M_Y'(0)M_T(0)}{M_Y(0)M_T'(0)} - c = \lambda\mu - c < 0 \tag{6.3}$$

by the net profit condition. Thus there exists at most a second solution to $\theta(R) = 0$ with $R > 0$. Again we call this solution the *adjustment coefficient* or the *Lundberg exponent*. Note that R is the strictly positive solution to the equation

$$M_Y(r)M_T(-cr) = 1\,.$$

Example 6.3 Let $\{C_t\}$ be a renewal risk model with Exp(1) distributed claim amounts, premium rate $c = 5$ and interarrival time distribution

$$F(t) = 1 - \frac{1}{2}(\mathrm{e}^{-3t} + \mathrm{e}^{-7t}).$$

It follows that $M_Y(r)$ exists for $r < 1$, $M_T(r)$ exists for $r < 3$ and $\lambda = 4.2$. The net profit condition $5 > 4.2$ is fulfilled. The equation to solve is

$$\frac{1}{1-r}\,\frac{1}{2}\Bigl(\frac{3}{3+5r} + \frac{7}{7+5r}\Bigr) = 1.$$

Thus

$$3(7+5r) + 7(3+5r) = 2(1-r)(3+5r)(7+5r)$$

or equivalently

$$25r^3 + 25r^2 - 4r = 0.$$

We find the obvious solution $r = 0$ and the two other solutions

$$r_{1/2} = \frac{-25 \pm \sqrt{1025}}{50} = \frac{-5 \pm \sqrt{41}}{10} = \begin{cases} 0.140312, \\ -1.14031. \end{cases}$$

From the theory we conclude that $R = 0.140312$. However, we proved that there is no negative solution. Why do we get $r_2 = -1.14031$? Obviously $M_Y(r_2) < 1 < \infty$, but $-cr_2 = 5.70156 > 3$ and thus $M_T(-cr_2) = \infty$. Thus r_2 is not a solution to the equation $M_Y(r)M_T(-cr) = 1$. ∎

Example 6.4 Let $T \sim \Gamma(\gamma, \alpha)$. Assume that there exists an r_∞ such that $M_Y(r) < \infty \iff r < r_\infty$. Then

$$\lim_{r \uparrow r_\infty} M_Y(r) = \infty.$$

Let $R(\gamma, \alpha)$ denote the Lundberg exponent, which exists in this case, and $\theta(r; \gamma, \alpha)$ denote the corresponding $\theta(r)$. Then for $h > 0$

$$\Bigl(\frac{\alpha}{\alpha + cR(\gamma,\alpha)}\Bigr)^{\gamma+h} M_Y(R(\gamma,\alpha)) = \Bigl(\frac{\alpha}{\alpha + cR(\gamma,\alpha)}\Bigr)^{h} < 1.$$

This means that $\theta(R(\gamma,\alpha); \gamma + h, \alpha) < 0$ and therefore $R(\gamma + h, \alpha) > R(\gamma, \alpha)$. Similarly

$$\Bigl(\frac{\alpha + h}{\alpha + h + cR(\gamma,\alpha)}\Bigr)^{\gamma} M_Y(R(\gamma,\alpha)) > \Bigl(\frac{\alpha}{\alpha + cR(\gamma,\alpha)}\Bigr)^{\gamma} M_Y(R(\gamma,\alpha)) = 1$$

and thus $R(\gamma, \alpha + h) < R(\gamma, \alpha)$. It would be interesting to see what happens if we let $\alpha \to 0$ and $\gamma \to 0$ in such a way that the expected value remains constant. Let therefore $\alpha = \lambda\gamma$. Recall that $m_Y(r) = \log M_Y(r)$.

Observe that

$$\lim_{\gamma \to 0} \gamma \log\Bigl(\frac{\lambda\gamma}{\lambda\gamma + cr}\Bigr) + m_Y(r) = m_Y(r)$$

and therefore $R(\gamma) = R(\gamma, \lambda\gamma) \to 0$ as $\gamma \to 0$. The latter follows because convergence to a continuous function is uniform on compact intervals. Let now $\rho(\gamma) = \gamma^{-1}R(\gamma)$. Then $\rho(\gamma)$ has to solve the equation

$$\log\Bigl(\frac{\lambda}{\lambda + c\rho}\Bigr) + \frac{m_Y(\gamma\rho)}{\gamma} = 0\,. \tag{6.4}$$

Letting $\gamma \to 0$ yields (in the limit ρ is an accumulation point)

$$\log\Bigl(\frac{\lambda}{\lambda + c\rho}\Bigr) + \rho m_Y'(0) = \log\Bigl(\frac{\lambda}{\lambda + c\rho}\Bigr) + \rho\mu = 0\,.$$

The latter function is convex in ρ, has the root $\rho = 0$, its derivative $-c/\lambda + \mu < 0$ at 0 and it diverges to infinity as $\rho \to \infty$. Thus there exists exactly one additional solution ρ_0 and $\rho(\gamma) \to \rho_0$ as $\gamma \to 0$.

Replace ρ by $\rho(\gamma)$ in (6.4) and take the derivative, which exists by the implicit function theorem,

$$-\frac{c\rho'(\gamma)}{\lambda + c\rho(\gamma)} + \frac{(\gamma\rho'(\gamma) + \rho(\gamma))m_Y'(\gamma\rho(\gamma))}{\gamma} - \frac{m_Y(\gamma\rho(\gamma))}{\gamma^2} = 0\,.$$

One obtains

$$\rho'(\gamma) = \frac{\gamma\rho(\gamma)m_Y'(\gamma\rho(\gamma)) - m_Y(\gamma\rho(\gamma))}{\gamma^2} \Big/ \Bigl(\frac{c}{\lambda + c\rho(\gamma)} - m_Y'(\gamma\rho(\gamma))\Bigr)\,.$$

Note that by Taylor expansion

$$\gamma\rho(\gamma)m_Y'(\gamma\rho(\gamma)) - m_Y(\gamma\rho(\gamma)) = \frac{\sigma^2}{2}\gamma^2\rho(\gamma)^2 + O(\gamma^3\rho(\gamma)^3) = \frac{\sigma^2}{2}\gamma^2\rho_0^2 + O(\gamma^3)$$

uniformly on compact intervals for which $R(\gamma)$ exists. The second equality holds because $\lim_{\gamma\downarrow 0}\rho(\gamma) = \rho_0 > 0$ exists. We denoted by σ^2 the variance of the claims. We found

$$\rho'(\gamma) = \frac{\sigma^2\rho_0^2(\lambda + c\rho_0)}{2(c - \lambda\mu - c\mu\rho_0)} + O(\gamma)\,.$$

Because the convergence is uniform it follows that

$$\rho(\gamma) = \rho_0 + \frac{\sigma^2 \rho_0^2(\lambda + c\rho_0)}{2(c - \lambda\mu - c\mu\rho_0)}\gamma + O(\gamma^2)$$

or

$$R(\gamma) = \gamma\rho_0 + \frac{\sigma^2 \rho_0^2(\lambda + c\rho_0)}{2(c - \lambda\mu - c\mu\rho_0)}\gamma^2 + O(\gamma^3)\,.$$

Note that the moment generating function of the interarrival times

$$\Bigl(\frac{\lambda\gamma}{\lambda\gamma - r}\Bigr)^{\gamma} = \Bigl(1 - \frac{r}{\lambda\gamma}\Bigr)^{-\gamma} \to \mathrm{e}^{r/\lambda}$$

converges to the moment generating function of deterministic interarrival times $T = 1/\lambda$. This means that $T(\gamma) \to 1/\lambda$ in distribution as $\gamma \to 0$, but the Lundberg exponents do not converge to the Lundberg exponent of the model with deterministic interarrival times. ■

6.3 Lundberg's Inequality

6.3.1 The Ordinary Case

In order to find an upper bound for the ruin probability we try to proceed as in the classical case. Unfortunately we cannot find an analogue of (5.1). However, recall the interpretation of (5.2) as conditioning on the ladder heights. Let $H(x) = \mathbb{P}[\tau < \infty, C_\tau \ge -x \mid C_0 = 0]$. Then

$$\psi(u) = \int_0^u \psi(u - x)\,\mathrm{d}H(x) + (H(\infty) - H(u))\,, \tag{6.5}$$

where $H(\infty) = \mathbb{P}[\tau < \infty \mid C_0 = 0]$. Unfortunately we cannot find an explicit expression for $H(u)$. In order to find a result we have to link the Lundberg exponent R to H.

Lemma 6.5 *Let $\{C_t\}$ be an ordinary renewal risk model with Lundberg exponent R. Then*

$$\int_0^\infty \mathrm{e}^{Rx}\,\mathrm{d}H(x) = \mathbb{E}[\mathrm{e}^{-RC_\tau} 1\!\!1_{\{\tau<\infty\}} \mid C_0 = 0] = 1\,.$$

Proof Let $B(x)$ be the weekly descending ladder height of the process $\{-C_{T_n}\}$. Multiplying the Wiener–Hopf-factorisation (E.1) by e^{rx} and integrating yields

$$M_Y(r)M_T(-cr) = M_H(r) + M_B(r) - M_H(r)M_B(r)\,.$$

We see that $M_H(r) < \infty$ iff $M_Y(r) < \infty$ and $M_B(r) < \infty$ iff $M_T(-cr) < \infty$. For $r = R$ the left-hand side is one and we can rewrite the equation as $(M_H(R) - 1)(1 - M_B(R)) = 0$. Since $B(x)$ is a distribution on the negative half line we have $M_B(R) < 1$. Thus $M_H(R) = 1$, which is the assertion. □

Here we can also give a martingale proof. The discrete time process $\{e^{-RC_{T_i}}\}$ is a martingale with initial value 1 if $C_0 = 0$. We will drop the conditioning on $C_0 = 0$ for the rest of the proof. By the stopping theorem

$$1 = \mathbb{E}[e^{-RC_{\tau\wedge T_n}}] = \mathbb{E}[e^{-RC_\tau} 1\!\!1_{\{\tau \le T_n\}}] + \mathbb{E}[e^{-RC_{T_n}} 1\!\!1_{\{\tau > T_n\}}] .$$

The function $e^{-RC_\tau} 1\!\!1_{\{\tau \le T_n\}}$ is increasing with n. By the monotone limit theorem it follows that

$$\lim_{n\to\infty} \mathbb{E}[e^{-RC_\tau} 1\!\!1_{\{\tau \le T_n\}}] = \mathbb{E}\Bigl[\lim_{n\to\infty} e^{-RC_\tau} 1\!\!1_{\{\tau \le T_n\}}\Bigr] = \mathbb{E}[e^{-RC_\tau} 1\!\!1_{\{\tau < \infty\}}] .$$

For the second term we have $e^{-RC_{T_n}} 1\!\!1_{\{\tau > T_n\}} \le 1$ and we know that $C_{T_n} \to \infty$ as $n \to \infty$. Thus

$$\lim_{n\to\infty} \mathbb{E}[e^{-RC_{T_n}} 1\!\!1_{\{\tau > T_n\}}] = \mathbb{E}\Bigl[\lim_{n\to\infty} e^{-RC_{T_n}} 1\!\!1_{\{\tau > T_n\}}\Bigr] = 0 .$$

We are now ready to prove Lundberg's inequality.

Theorem 6.6 *Let $\{C_t\}$ be an ordinary renewal risk model and assume that the Lundberg exponent R exists. Then*

$$\psi(u) < e^{-Ru} .$$

Proof Assume that the assertion is wrong and let

$$u_0 = \inf\{u \ge 0 : \psi(u) \ge e^{-Ru}\} .$$

Let $u_0 \le u < u_0 + \varepsilon$ such that $\psi(u) \ge e^{-Ru}$. Then

$$\psi(u_0) \ge \psi(u) \ge e^{-Ru} > e^{-R(u_0+\varepsilon)} .$$

Thus $\psi(u_0) \ge e^{-Ru_0}$. Because $e^{-RC_\tau} > 1$ it follows from Lemma 6.5 that $\psi(0) = H(\infty) < 1$ and therefore $u_0 \ge -R^{-1} \log H(\infty) > 0$. Assume first that $H(u_0-) = 0$. Then it follows from (6.5) that

$$\begin{aligned} e^{-Ru_0} &\le \psi(u_0) = \psi(0)H(u_0) + \int_{u_0}^\infty dH(x) < H(u_0) + \int_{u_0}^\infty e^{R(x-u_0)}\, dH(x) \\ &= \int_{u_0-}^\infty e^{R(x-u_0)}\, dH(x) = e^{-Ru_0} \int_0^\infty e^{Rx}\, dH(x) = e^{-Ru_0} , \end{aligned}$$

which is a contradiction. Thus $H(u_0-) > 0$.

As in the classical case we obtain from (6.5) and Lemma 6.5

$$\begin{aligned}
\mathrm{e}^{-Ru_0} \le \psi(u_0) &= \int_0^{u_0} \psi(u_0 - x)\,\mathrm{d}H(x) + \int_{u_0}^{\infty} \mathrm{d}H(x) \\
&< \int_0^{u_0} \mathrm{e}^{-R(u_0-x)}\,\mathrm{d}H(x) + \int_{u_0}^{\infty} \mathrm{d}H(x) \\
&\le \int_0^{\infty} \mathrm{e}^{-R(u_0-x)}\,\mathrm{d}H(x) = \mathrm{e}^{-Ru_0}\,.
\end{aligned}$$

This is a contradiction and the theorem is proved. □

As in the classical case we prove the result using martingale techniques. By the stopping theorem

$$\begin{aligned}
\mathrm{e}^{-Ru} &= \mathrm{e}^{-RC_{T_0}} = \mathbb{E}[\mathrm{e}^{-RC_{\tau\wedge T_n}}] \ge \mathbb{E}[\mathrm{e}^{-RC_{\tau\wedge T_n}}; \tau \le T_n] \\
&= \mathbb{E}[\mathrm{e}^{-RC_\tau}; \tau \le T_n]\,.
\end{aligned}$$

Letting $n \to \infty$ yields by monotone convergence

$$\mathrm{e}^{-Ru} \ge \mathbb{E}[\mathrm{e}^{-RC_\tau}; \tau < \infty] > \mathbb{P}[\tau < \infty]$$

because $C_\tau < 0$.

Example 6.7 Let Y_i be Exp(α) distributed and

$$F(t) = 1 - p\mathrm{e}^{-\beta t} - (1-p)\mathrm{e}^{-\gamma t}\,,$$

where $0 < \beta < \gamma$ and $0 < p \le \gamma(\gamma - \beta)^{-1}$. The net profit condition can then be written as

$$\alpha c(\beta + p(\gamma - \beta)) > \beta\gamma\,.$$

The equation for the adjustment coefficient is

$$\frac{\alpha}{\alpha - r}\Bigl(\frac{p\beta}{\beta + cr} + \frac{(1-p)\gamma}{\gamma + cr}\Bigr) = 1\,.$$

This is equivalent to

$$c^2 r^3 - c(\alpha c - \beta - \gamma)r^2 - (\alpha c(\beta + p(\gamma - \beta)) - \beta\gamma)r = 0$$

from which the solutions $r_0 = 0$ and

$$r_{1/2} = \frac{\alpha c - \beta - \gamma \pm \sqrt{(\alpha c - \gamma + \beta)^2 + 4\alpha c p(\gamma - \beta)}}{2c}$$

follow. Note that by the net profit condition $r_1 > 0 > r_2$. By the condition on p

$$r_2 < r_1 \le \frac{\alpha c - \beta - \gamma + \sqrt{(\alpha c - \gamma + \beta)^2 + 4\alpha c\gamma}}{2c} < \frac{\alpha c - \beta - \gamma + \sqrt{(\alpha c + \gamma + \beta)^2}}{2c} = \alpha\,.$$

Thus $M_Y(r_{1/2}) < \infty$. We have

$$-cr_2 = \frac{\beta + \gamma - \alpha c + \sqrt{(\alpha c - \gamma + \beta)^2 + 4\alpha cp(\gamma - \beta)}}{2} > \frac{\beta + \gamma - \alpha c + \sqrt{(\alpha c - \gamma + \beta)^2}}{2} \ge \beta$$

and thus $M_T(-cr_2) = \infty$, i.e. r_2 is not a solution. Therefore

$$R = \frac{\alpha c - \beta - \gamma + \sqrt{(\alpha c - \gamma + \beta)^2 + 4\alpha cp(\gamma - \beta)}}{2c}\,.$$

It follows from Lundberg's inequality that

$$\psi(u) < \mathrm{e}^{-Ru}\,.$$

■

6.3.2 The General Case

Now let F^1 be arbitrary. Let us denote the ruin probability in the ordinary model by $\psi^{\mathrm{o}}(u)$. We know that $\{C_{T_1+t}\}$ is an ordinary model with initial capital C_{T_1}. There are two possibilities: Ruin occurs at T_1 or $C_{T_1} \ge 0$. Thus

$$\begin{aligned}
\psi(u) &= \int_0^\infty \Bigl(\int_0^{u+ct} \psi^{\mathrm{o}}(u + ct - y)\,\mathrm{d}G(y) + \int_{u+ct}^\infty \mathrm{d}G(y)\Bigr)\,\mathrm{d}F^1(t) \\
&< \int_0^\infty \Bigl(\int_0^{u+ct} \mathrm{e}^{-R(u+ct-y)}\,\mathrm{d}G(y) + \int_{u+ct}^\infty \mathrm{e}^{-R(u+ct-y)}\,\mathrm{d}G(y)\Bigr)\,\mathrm{d}F^1(t) \\
&= \int_0^\infty \int_0^\infty \mathrm{e}^{-R(u+ct-y)}\,\mathrm{d}G(y)\,\mathrm{d}F^1(t) \\
&= \mathrm{e}^{-Ru}\mathbb{E}[\mathrm{e}^{R(Y_1 - cT_1)}] = M_Y(R)M_{T_1}(-cR)\mathrm{e}^{-Ru}\,.
\end{aligned}$$

Thus $\psi(u) < C\mathrm{e}^{-Ru}$ for $C = M_Y(R)M_{T_1}(-cR)$. In the cases considered so far we always had $C = 1$, but in the general case $C \in (0, M_Y(R))$ can be any value in this interval (let T_1 be deterministic). If $F^1(t) = F(t)$ then $C = 1$ by the equation determining the adjustment coefficient.

6.4 The Cramér–Lundberg Approximation

6.4.1 The Ordinary Case

In order to obtain the Cramér–Lundberg approximation we proceed as in the classical case and multiply (6.5) by e^{Ru}. We have

$$\psi(u)e^{Ru} = \int_0^u \psi(u-x)e^{R(u-x)}e^{Rx}\,dH(x) + e^{Ru}(H(\infty) - H(u))\,.$$

We obtain the following

Theorem 6.8 *Let $\{C_t\}$ be an ordinary renewal risk model. Assume that R exists and that there exists an $r > R$ such that $M_Y(r) < \infty$. If the distribution of $Y_1 - cT_1$ given $Y_1 - cT_1 > 0$ is not arithmetic then*

$$\lim_{u\to\infty} \psi(u)e^{Ru} = \frac{1 - H(\infty)}{R\int_0^\infty xe^{Rx}\,dH(x)} =: C\,.$$

If the distribution of $Y_1 - cT_1$ is arithmetic with span γ then for $x \in [0, \gamma)$

$$\lim_{n\to\infty} \psi(x + n\gamma)e^{R(x+n\gamma)} = Ce^{Rx}\frac{1 - e^{-R\gamma}}{R}\,.$$

Proof We prove only the non-arithmetic case. The arithmetic case can be proved similarly. Recall from Lemma 6.5 that $\int_0^x e^{Ry}\,dH(y)$ is a proper distribution function. It follows from Lemma C.14 that $e^{Ru}(H(\infty) - H(u))$ is directly Riemann integrable. From $M_Y(r) < \infty$ for $r > R$ it follows that $\int_0^\infty xe^{Rx}\,dH(x) < \infty$. Thus by the key renewal theorem

$$\lim_{u\to\infty} \psi(u)e^{Ru} = \frac{\int_0^\infty e^{Ru}(H(\infty) - H(u))\,du}{\int_0^\infty xe^{Rx}\,dH(x)}\,.$$

Simplifying the numerator we find

$$\int_0^\infty e^{Ru}\int_u^\infty dH(x)\,du = \int_0^\infty\int_0^x e^{Ru}\,du\,dH(x) = \frac{1}{R}(1 - H(\infty))\,,$$

where we used Lemma 6.5 in the last equality. □

It should be remarked that in general there is no explicit expression for the ladder height distribution $H(x)$. It is therefore difficult to find an explicit expression for C. Nevertheless we know that $0 < C \le 1$.

Assume that we know C. Then

$$\psi(u) \approx C\mathrm{e}^{-Ru}$$

for u large.

Example 6.9 Assume that Y_i is Exp(α) distributed. Because $M_Y(r) \to \infty$ as $r \to \alpha$ it follows that R exists and that the conditions of Theorem 6.8 are fulfilled. Consider the martingale $\{\mathrm{e}^{-RC_{T_i}}\}$. By the stopping theorem

$$\mathrm{e}^{-Ru} = \mathbb{E}[\mathrm{e}^{-RC_{\tau\wedge T_i}}] = \mathbb{E}[\mathrm{e}^{-RC_\tau} 1\!\!1_{\{\tau \le T_i\}}] + \mathbb{E}[\mathrm{e}^{-RC_{T_i}} 1\!\!1_{\{\tau > T_i\}}] \,.$$

As in the proof of Lemma 6.5 it follows that

$$\mathrm{e}^{-Ru} = \mathbb{E}[\mathrm{e}^{-RC_\tau} 1\!\!1_{\{\tau < \infty\}}] = \mathbb{E}[\mathrm{e}^{-RC_\tau} \mid \tau < \infty]\psi(u) \,.$$

Assume for the moment that we know $C_{\tau-}$. Let $Z = C_{\tau-} - C_\tau$ denote the size of the claim leading to ruin. The only information on Z we have is that $Z > C_{\tau-}$ because ruin occurs at time τ. Then

$$\begin{aligned}\mathbb{P}[-C_\tau > x \mid C_{\tau-} = y, \tau < \infty] &= \mathbb{P}[Z > y + x \mid C_{\tau-} = y, \tau < \infty] \\ &= \mathbb{P}[Y_1 > y + x \mid Y_1 > y] = \mathrm{e}^{-\alpha x} \,.\end{aligned}$$

It follows that

$$\mathbb{E}[\mathrm{e}^{-RC_\tau} \mid \tau < \infty] = \int_0^\infty \mathrm{e}^{Rx} \alpha \mathrm{e}^{-\alpha x} \,\mathrm{d}x = \frac{\alpha}{\alpha - R} \,.$$

Thus we obtain the explicit solution

$$\psi(u) = \frac{\alpha - R}{\alpha} \mathrm{e}^{-Ru}$$

to the ruin problem. Therefore, as in the classical case, the Cramér–Lundberg approximation is exact and we have an explicit expression for C. ■

6.4.2 The General Case

Consider the equation

$$\psi(u) = \int_0^\infty \Bigl(\int_0^{u+ct} \psi^{\text{o}}(u+ct-y)\,\mathrm{d}G(y) + (1-G(u+ct))\Bigr)\,\mathrm{d}F^1(t)\,.$$

We have to multiply the above equation by e^{Ru}. We obtain by Lundberg's inequality

$$\begin{aligned}
&\int_0^\infty \int_0^{u+ct} \psi^{\text{o}}(u+ct-y)\mathrm{e}^{Ru}\,\mathrm{d}G(y)\,\mathrm{d}F^1(t) \\
&\quad= \int_0^\infty \int_0^{u+ct} \psi^{\text{o}}(u+ct-y)\mathrm{e}^{R(u+ct-y)}\mathrm{e}^{Ry}\,\mathrm{d}G(y)\mathrm{e}^{-cRt}\,\mathrm{d}F^1(t) \\
&\quad< \int_0^\infty \int_0^{u+ct} \mathrm{e}^{Ry}\,\mathrm{d}G(y)\mathrm{e}^{-cRt}\,\mathrm{d}F^1(t) \\
&\quad\le \int_0^\infty \int_0^\infty \mathrm{e}^{Ry}\,\mathrm{d}G(y)\mathrm{e}^{-cRt}\,\mathrm{d}F^1(t) = M_Y(R)M_{T_1}(-cR) < \infty\,. \qquad (6.6)
\end{aligned}$$

We can therefore interchange limit and integration

$$\begin{aligned}
&\lim_{u\to\infty} \int_0^\infty \int_0^{u+ct} \psi^{\text{o}}(u+ct-y)\mathrm{e}^{R(u+ct-y)}\mathrm{e}^{Ry}\,\mathrm{d}G(y)\mathrm{e}^{-cRt}\,\mathrm{d}F^1(t) \\
&\quad= \int_0^\infty \int_0^\infty C\mathrm{e}^{Ry}\,\mathrm{d}G(y)\mathrm{e}^{-cRt}\,\mathrm{d}F^1(t) = M_Y(R)M_{T_1}(-cR)C\,.
\end{aligned}$$

The remaining term converges to 0 because

$$\begin{aligned}
\int_0^\infty \int_{u+ct}^\infty \mathrm{e}^{Ru}\,\mathrm{d}G(y)\,\mathrm{d}F^1(t) &\le \int_0^\infty \int_{u+ct}^\infty \mathrm{e}^{R(y-ct)}\,\mathrm{d}G(y)\,\mathrm{d}F^1(t) \\
&\le \int_0^\infty \int_u^\infty \mathrm{e}^{Ry}\,\mathrm{d}G(y)\,\mathrm{e}^{-cRt}\,\mathrm{d}F^1(t) \\
&= M_{T_1}(-cR) \int_u^\infty \mathrm{e}^{Ry}\,\mathrm{d}G(y)\,.
\end{aligned}$$

Hence

$$\lim_{u\to\infty} \psi(u)\mathrm{e}^{Ru} = M_Y(R)M_{T_1}(-cR)C\,, \qquad (6.7)$$

where C was the constant obtained in the ordinary case.

Example 6.10 Consider again exponentially distributed claim sizes. In the general case we have to change the martingale because in general $\mathbb{E}[\mathrm{e}^{-RC_{T_1}}] \ne \mathrm{e}^{-Ru}$. Note that

$$\mathbb{E}[\mathrm{e}^{-RC_{T_1}}] = \mathrm{e}^{-Ru}M_Y(R)M_{T_1}(-cR) = \frac{\alpha}{\alpha-R}\mathrm{e}^{-Ru}M_{T_1}(-cR)\,.$$

Let

$$M_n = \begin{cases} \mathrm{e}^{-RC_{T_n}} & \text{if } n \ge 1, \\ \frac{\alpha}{\alpha-R}\mathrm{e}^{-Ru}M_{T_1}(-cR) & \text{if } n = 0. \end{cases}$$

As in the ordinary case we obtain

$$\frac{\alpha}{\alpha - R} e^{-Ru} M_{T_1}(-cR) = \frac{\alpha}{\alpha - R} \psi(u)$$

or equivalently

$$\psi(u) = M_{T_1}(-cR) e^{-Ru} .$$

■

6.5 Diffusion Approximations

As in the classical case it is possible to prove the following

Proposition 6.11 *Let $\{C_t^{(n)}\}$ be a sequence of renewal risk models with initial capital u, interarrival time distribution $F^{(n)}(t) = F(nt)$, claim size distribution $G^{(n)}(x) = G(x\sqrt{n})$ and premium rate*

$$c^{(n)} = \Bigl(1 + \frac{\frac{c-\lambda\mu}{\lambda\mu}}{\sqrt{n}}\Bigr)\lambda^{(n)}\mu^{(n)} = c + (\sqrt{n} - 1)\lambda\mu$$

where $\lambda = \lambda^{(1)}$ and $\mu = \mu^{(1)}$. Then

$$\{C_t^{(n)}\} \to \{u + W_t\}$$

uniformly on finite intervals where $\{W_t\}$ is a $(c - \lambda\mu, \lambda\mu_2)$-Brownian motion.

Proof See, for instance, [80, p. 172]. □

Denote the ruin time of the renewal model by $\tau^{(n)}$ and the ruin time of the Brownian motion $\tau = \inf\{t \ge 0 : u + W_t < 0\}$. The diffusion approximation is based on the following

Proposition 6.12 *Let $\{C_t^{(n)}\}$ and $\{W_t\}$ as above. Then*

$$\lim_{n\to\infty} \mathbb{P}[\tau^{(n)} \le t] = \mathbb{P}[\tau \le t] .$$

Proof See [138, Thm. 9]. □

As in the classical case numerical investigations show that the approximation only works well if $c/(\lambda\mu)$ is close to 1.

6.6 Subexponential Claim Size Distributions

In the classical case we had assumed that the ladder height distribution is subexponential. If we assume that the ladder height distribution H is subexponential then the asymptotic behaviour would follow as in the classical case. Unfortunately we do not have an explicit expression for the ladder height distribution in the renewal case. The following proposition helps us out of this problem. We assume the ordinary case.

Proposition 6.13 *Let U be the distribution of $u - C_{T_1} = Y_1 - cT_1$ and let $m = \int_0^\infty (1 - U(x))\,\mathrm{d}x$. Let $G_1(x) = \mu^{-1}\int_0^x (1 - G(y))\,\mathrm{d}y$ and $U_1(x) = m^{-1}\int_0^x (1 - U(y))\,\mathrm{d}y$. Then*

(i) If G_1 is subexponential, then U_1 is subexponential and

$$\lim_{x\to\infty} \frac{1 - U_1(x)}{1 - G_1(x)} = \frac{\mu}{m}\,.$$

(ii) If U_1 is subexponential then $H/H(\infty)$ is subexponential and

$$\lim_{x\to\infty} \frac{1 - U_1(x)}{H(\infty) - H(x)} = \frac{c - \lambda\mu}{\lambda m(1 - H(\infty))}\,.$$

Proof (i) Using Fubini's theorem

$$\begin{aligned} m(1 - U_1(x)) &= \int_x^\infty \int_0^\infty (1 - G(y + ct))\,\mathrm{d}F(t)\,\mathrm{d}y \\ &= \int_0^\infty \int_x^\infty (1 - G(y + ct))\,\mathrm{d}y\,\mathrm{d}F(t) \\ &= \mu \int_0^\infty (1 - G_1(x + ct))\,\mathrm{d}F(t)\,. \end{aligned}$$

It follows by the bounded convergence theorem and Lemma F.2 that

$$\lim_{x\to\infty} \frac{1 - U_1(x)}{1 - G_1(x)} = \lim_{x\to\infty} \frac{\mu}{m} \int_0^\infty \frac{1 - G_1(x + ct)}{1 - G_1(x)}\,\mathrm{d}F(t) = \frac{\mu}{m}\,.$$

By Lemma F.8 U_1 is subexponential.
(ii) Consider the random walk $\sum_{i=1}^n (Y_i - cT_i)$. We use the notation of Lemma E.2. By the Wiener–Hopf factorisation for $y > 0$

$$1 - U(y) = \int_{-\infty}^0 (H(y - z) - H(y))\,\mathrm{d}\rho(z)\,.$$

Let $0 < x < b$. Then using Fubini's theorem

$$\int_x^b (1 - U(y))\,\mathrm{d}y = \int_{-\infty}^0 \int_x^b (H(y - z) - H(y))\,\mathrm{d}y\,\mathrm{d}\rho(z)$$
$$= \int_{-\infty}^0 \Bigl(\int_x^{x-z} (H(\infty) - H(y))\,\mathrm{d}y - \int_b^{b-z} (H(\infty) - H(y))\,\mathrm{d}y\Bigr)\,\mathrm{d}\rho(z)\,.$$

Because $\int_{-\infty}^0 |z|\,\mathrm{d}\rho(z)$ is a finite upper bound (Lemma E.2) we can interchange the integral and the limit $b \to \infty$ and obtain

$$\int_x^\infty (1 - U(y))\,\mathrm{d}y = \int_{-\infty}^0 \int_x^{x-z} (H(\infty) - H(y))\,\mathrm{d}y\,\mathrm{d}\rho(z)\,.$$

We find

$$\int_{-\infty}^0 |z|(H(\infty) - H(x - z))\,\mathrm{d}\rho(z) \le m(1 - U_1(x))$$
$$\le (H(\infty) - H(x)) \int_{-\infty}^0 |z|\,\mathrm{d}\rho(z)\,.$$

For $s > 0$ we obtain

$$(H(\infty) - H(x + s)) \int_{-s}^0 |z|\,\mathrm{d}\rho(z) \le \int_{-s}^0 |z|(H(\infty) - H(x - z))\,\mathrm{d}\rho(z)$$
$$\le m(1 - U_1(x))$$

and therefore

$$1 \le \int_{-\infty}^0 |z|\,\mathrm{d}\rho(z) \frac{H(\infty) - H(x + s)}{m(1 - U_1(x + s))} \le \frac{\int_{-\infty}^0 |z|\,\mathrm{d}\rho(z)}{\int_{-s}^0 |z|\,\mathrm{d}\rho(z)} \frac{1 - U_1(x)}{1 - U_1(x + s)}\,.$$

Thus by Lemma F.2 $(H(\infty) - H(x))^{-1}(1 - U_1(x)) \to m^{-1} \int_{-\infty}^0 |z|\,\mathrm{d}\rho(z)$ and $H/H(\infty)$ is subexponential by Lemma F.8. The explicit value of the limit follows from Lemma E.2, noting that $\int_{-\infty}^0 |z|\,\mathrm{d}\rho(z)$ is the expected value of the first descending ladder height of the random walk $\sum_{i=1}^n (Y_i - cT_i)$. □

We can now prov e the asymptotic behaviour of the ruin probability.

Theorem 6.14 *Let $\{C_t\}$ be a renewal risk process. Assume that the integrated tail of the claim size distribution*

$$G_1(x) = \mu^{-1} \int_0^x (1 - G(y))\,\mathrm{d}y$$

is subexponential. Then

$$\lim_{u\to\infty} \frac{\psi(u)}{\int_u^\infty (1-G(y))\,\mathrm{d}y} = \frac{\lambda}{c-\lambda\mu}\;.$$

Remark 6.15 Note that this is exactly Theorem 5.26. The difference is only that $G_1(x)$ is no longer the ladder height distribution. ■

Proof Let us first consider an ordinary renewal model. By repeating the proof of Theorem 5.26 we find that

$$\lim_{u\to\infty} \frac{\psi(u)}{1-H(u)/H(\infty)} = \frac{H(\infty)}{1-H(\infty)}$$

because $H(x)/H(\infty)$ is subexponential by Proposition 6.13. Moreover, by Proposition 6.13

$$\begin{aligned}
&\lim_{u\to\infty} \frac{\psi(u)}{\int_u^\infty (1-G(y))\,\mathrm{d}y} \\
&\quad = \lim_{u\to\infty} \frac{\psi(u)}{1-H(u)/H(\infty)} \frac{1}{H(\infty)} \frac{H(\infty)-H(u)}{1-U_1(u)} \frac{1-U_1(u)}{\mu(1-G_1(u))} \\
&\quad = \frac{H(\infty)}{1-H(\infty)} \frac{1}{H(\infty)} \frac{\lambda m(1-H(\infty))}{c-\lambda\mu} \frac{1}{m} = \frac{\lambda}{c-\lambda\mu}\;.
\end{aligned}$$

For an arbitrary renewal risk model the assertion follows similarly as in Sect. 6.4.2. However, two properties of subexponential distributions that are not proved here will be needed. □

6.7 Finite Time Lundberg Inequalities

Let $0 \le \underline{y} < \overline{y} < \infty$ and define $T = \inf\{T_n : n \in \mathbb{N}, T_n \ge \overline{y}u\}$. Using the stopping theorem

$$\begin{aligned}
\mathrm{e}^{-ru} &= \mathbb{E}[\mathrm{e}^{-rC_{\tau\wedge T\wedge T_n}-\theta(r)(\tau\wedge T\wedge T_n)}] \\
&= \mathbb{E}[\mathrm{e}^{-rC_{\tau\wedge T}-\theta(r)(\tau\wedge T)}; T_n \ge \tau\wedge T] + \mathbb{E}[\mathrm{e}^{-rC_{T_n}-\theta(r)T_n}; T_n < \tau\wedge T]\;.
\end{aligned}$$

By monotone convergence

$$\lim_{n\to\infty} \mathbb{E}[\mathrm{e}^{-rC_{\tau\wedge T}-\theta(r)(\tau\wedge T)}; T_n \ge \tau\wedge T] = \mathbb{E}[\mathrm{e}^{-rC_{\tau\wedge T}-\theta(r)(\tau\wedge T)}]\;.$$

For the second term we obtain

$$\lim_{n\to\infty} \mathbb{E}[\mathrm{e}^{-rC_{T_n}-\theta(r)T_n}; T_n < \tau\wedge T] \le \lim_{n\to\infty} \max\{\mathrm{e}^{-\theta(r)\overline{y}u}, 1\}\mathbb{P}[T_n < \tau\wedge T] = 0\;.$$

Therefore

$$\begin{aligned}
e^{-ru} &= \mathbb{E}[e^{-rC_{\tau\wedge T}-\theta(r)(\tau\wedge T)}] > \mathbb{E}[e^{-rC_{\tau\wedge T}-\theta(r)(\tau\wedge T)}; \underline{y}u < \tau \le \overline{y}u] \\
&= \mathbb{E}[e^{-rC_{\tau}-\theta(r)\tau} \mid \underline{y}u < \tau \le \overline{y}u]\mathbb{P}[\underline{y}u < \tau \le \overline{y}u] \\
&> \mathbb{E}[e^{-\theta(r)\tau} \mid \underline{y}u < \tau \le \overline{y}u]\mathbb{P}[\underline{y}u < \tau \le \overline{y}u] \\
&> e^{-\max\{\theta(r)\underline{y}u,\theta(r)\overline{y}u\}}\mathbb{P}[\underline{y}u < \tau \le \overline{y}u] .
\end{aligned}$$

We obtain

$$\mathbb{P}[\underline{y}u < \tau \le \overline{y}u] < e^{-(r-\max\{\theta(r)\underline{y},\theta(r)\overline{y}\})u} = e^{-\min\{r-\theta(r)\underline{y},r-\theta(r)\overline{y}\}u} ,$$

which leads to the finite time Lundberg inequality

$$\mathbb{P}[\underline{y}u < \tau \le \overline{y}u] < e^{-R(\underline{y},\overline{y})u} .$$

Here, as in the classical case,

$$R(\underline{y},\overline{y}) := \sup\{\min\{r-\theta(r)\underline{y}, r-\theta(r)\overline{y}\} : r \in \mathbb{R}\} .$$

We have already discussed $R(\underline{y},\overline{y})$ in the classical case (Sect. 5.14). Assume that $r_\infty = \sup\{r \ge 0 : M_Y(r) < \infty\} > R$. Let

$$y_0 = (\theta'(R))^{-1} = \Bigl(\frac{M_Y'(R)M_T(-cR)}{M_Y(R)M_T'(-cR)} - c\Bigr)^{-1}$$

denote the critical value. Then we obtain as in the classical case

$$\mathbb{P}[0 < \tau \le yu \mid C_0 = u] < e^{-R(0,y)u}$$

and $R(0, y) > R$ if $y < y_0$, and

$$\mathbb{P}[yu < \tau < \infty \mid C_0 = u] < e^{-R(y,\infty)u}$$

and $R(y,\infty) > R$ if $y > y_0$. We can prove this in the same way as Theorem 5.38.

Theorem 6.16 *Assume that $R < r_\infty$. Then*

$$\frac{\tau}{u} \longrightarrow y_0$$

in probability on the set $\{\tau < \infty\}$. □

Bibliographical Remarks

The renewal risk model was introduced by Andersen [3]. The papers by Thorin [129–131] are also good references. Example 6.4 is due to Grandell [74, p. 71]. Lundberg's inequality in the ordinary case was first proved by Andersen [3, p. 224]. Example 6.7 can be found in [74, p. 60]. The Cramér–Lundberg approximation was first proved by Thorin, [129, p. 94] in the ordinary case and [130, p. 97] in the stationary case. The results on subexponential claim sizes go back to Embrechts and Veraverbeke [56]. The results on finite time Lundberg inequalities can be found in [55]. More results on the renewal risk model may be found, for example, in [11, 74, 110].

Chapter 7
The Ammeter Risk Model

In the same way as the negative binomial distribution is obtained by mixing the Poisson parameter with a Gamma distribution, we would like to mix the Poisson parameter into the classical risk model. Since, with time, the Poisson parameter is determined more and more precisely by the accumulation of new information, we repeat the mixing independently after a certain time period. We can prove again the basic results for the corresponding ruin probability.

7.1 Mixed Poisson Risk Processes

We have seen that a continuous time risk model is "easy" to handle if the claim number process is a Poisson process. Considering a risk in a fixed time interval we learned that a compound Poisson model does not fit real data very well. We had to use a compound mixed Poisson model such as, for instance, the compound negative binomial model. The reason that these models are closer to reality is that they allow more fluctuation.

It is natural to ask whether we can allow for more fluctuation in the Poisson process. We already saw one way to allow for more fluctuation, namely the renewal risk model, but this is not an analogue of the compound mixed Poisson model in a fixed time interval. The easiest way to obtain a mixed Poisson distribution for the number of claims in any finite time interval is to let the claim arrival intensity to be stochastic. Such a model is called a *mixed Poisson risk process*.

H. Schmidli, *Risk Theory*, Springer Actuarial,
DOI:10.1007/978-3-319-72005-0_7

Let $H(x)$ be the distribution function of λ. Given λ, the risk process is then the classical risk model with claim intensity λ. Then, using the usual notation,

$$\psi(u) = \int_0^\infty \psi^{\mathrm{C}}(u, \ell)\,\mathrm{d}H(\ell) = \int_0^{c/\mu} \psi^{\mathrm{C}}(u, \ell)\,\mathrm{d}H(\ell) + (1 - H(c/\mu))$$

where $\psi^{\mathrm{C}}(u, \ell)$ denotes the ruin probability in the Cramér–Lundberg model with claim arrival intensity ℓ. It follows immediately that there can only exist an exponential upper bound if there exists an ℓ_0 such that $H(\ell_0) = 1$ and $c > \mu\ell_0$. Otherwise, for instance if λ is gamma distributed, Lundberg's inequality or the Cramér–Lundberg approximation cannot hold. In particular, there is no way to formulate a net profit condition without knowing an upper bound for λ.

Assume we have observed the process up to time t. We want to show that $\mathbb{P}[\lambda \le \ell \mid \mathcal{F}_t] = \mathbb{P}[\lambda \le \ell \mid N_t]$. Because $\{N_t\}$ and $\{Y_i\}$ are independent it follows that $\mathbb{P}[\lambda \le \ell \mid \mathcal{F}_t] = \mathbb{P}[\lambda \le \ell \mid N_t, T_1, T_2, \ldots, T_{N_t}]$. Recall that, given $N_t = n$ the claim times $T_1, T_2, \ldots, T_n$ have the same distribution as the order statistics of n independent random variables uniformly distributed in $[0, t]$ (Proposition C.3). Let $n \in \mathbb{N} \setminus \{0\}$ and let $0 \le t_i \le t$ for $1 \le i \le n$. Then

$$\begin{aligned}
\mathbb{P}[\lambda \le{}&\ell, N_t = n, T_1 \le t_1, \ldots, T_n \le t_n] \\
&= \mathbb{P}[\lambda \le \ell, T_1 \le t_1, \ldots, T_n \le t_n, T_{n+1} > t] \\
&= \int_0^\ell \int_0^{t_1} x\mathrm{e}^{-xs_1} \int_{s_1}^{t_2 \vee s_1} x\mathrm{e}^{-x(s_2-s_1)} \cdots \int_{s_{n-1}}^{t_n \vee s_{n-1}} x\mathrm{e}^{-x(s_n-s_{n-1})} \\
&\qquad\qquad \times \mathrm{e}^{-x(t-s_n)}\,\mathrm{d}s_n \cdots \mathrm{d}s_1\,\mathrm{d}H(x) \\
&= \int_0^\ell \frac{(xt)^n}{n!}\mathrm{e}^{-xt} \int_0^{t_1} \int_{s_1}^{t_2 \vee s_1} \cdots \int_{s_{n-1}}^{t_n \vee s_{n-1}} n! t^{-n}\,\mathrm{d}s_n \ldots \mathrm{d}s_1\,\mathrm{d}H(x) \\
&= \mathbb{P}[\lambda \le \ell, N_t = n]\mathbb{P}[T_1 \le t_1, \ldots, T_n \le t_n \mid N_t = n] \\
&= \mathbb{P}[\lambda \le \ell \mid N_t = n]\mathbb{P}[T_1 \le t_1, \ldots, T_n \le t_n, N_t = n]
\end{aligned}$$

and it follows that

$$\mathbb{P}[\lambda \le \ell \mid N_t = n, T_1 \le t_1, \ldots, T_n \le t_n] = \mathbb{P}[\lambda \le \ell \mid N_t = n]\,.$$

Thus $\mathbb{P}[\lambda \le \ell \mid \mathcal{F}_t] = \mathbb{P}[\lambda \le \ell \mid N_t]$.

We know from renewal theory (or using the strong law of large numbers) that, given λ, $t^{-1}N_t$ converges to λ a.s. Thus

$$\mathbb{P}[\lim_{t\to\infty} t^{-1}N_t = \lambda] = 1\,.$$

Moreover, $\mathrm{Var}[\lambda \mid N_t] \to 0$ as $t \to \infty$. Thus as time increases the desired variability disappears. Hence this model is not the desired generalisation of the mixed Poisson model in a finite time interval.

7.2 Definition of the Ammeter Risk Model

In 1948 the Swiss actuary Hans Ammeter had the idea to combine a sequence of mixed Poisson models. He wanted to start a new independent model at the beginning of each year. That means a new intensity level is chosen at the beginning of each year independently of the past intensity levels. This has the advantage that the variability does not disappear with time.

For a formal definition let Δ denote the length of the time interval in which the intensity level remains constant, for instance one year. Let $\{L_i\}$ be a sequence of positive random variables with distribution function H. We allow for the possibility that $H(0) \neq 0$. The intensity at time t is

$$\lambda_t := L_i, \qquad \text{if } (i-1)\Delta \le t < i\Delta.$$

The stochastic process $\{\lambda_t\}$ is called the *claim arrival intensity process* of the risk model. Define $\Lambda_t = \int_0^t \lambda_s \, \mathrm{d}s$ and let $\{\tilde{N}_t\}$ be a homogeneous Poisson process with rate 1. The claim number process is then

$$N_t := \tilde{N}_{\Lambda_t} .$$

Intuitively the point process $\{N_t\}$ can be obtained in the following way. First generate a realisation $\{\lambda_t\}$. Given $\{\lambda_t : 0 \le t < \infty\}$ let $\{N_t\}$ be an inhomogeneous Poisson process with rate $\{\lambda_t\}$ (compare with Proposition C.7). A process obtained in this way is called a *doubly stochastic point process* or *Cox process*. The process $\{N_t\}$ constructed above and the mixed Poisson process are special cases of Cox processes.

As for the models considered before, let the claims $\{Y_i\}$ be a sequence of iid. positive random variables with distribution function G. The Ammeter risk model is the process

$$C_t = u + ct - \sum_{i=1}^{N_t} Y_i$$

where u is the initial capital and c is the premium rate. As before let $\tau := \inf\{t > 0 : C_t < 0\}$ denote the time of ruin and $\psi(u) = \mathbb{P}[\tau < \infty \mid C_0 = u]$ denote the ruin probability.

In order to investigate the model we also define the probability of ruin obtained by considering only the times $\{\Delta, 2\Delta, \ldots\}$ where the intensity changes. That is, let $\tau_\Delta = \inf\{i\Delta : i \in \mathbb{N}, C_{i\Delta} < 0\}$ and $\psi_\Delta(u) := \mathbb{P}[\tau_\Delta < \infty]$. By only considering the points $i\Delta$ we are essentially considering a model in which all the claims of the interval $((i-1)\Delta, i\Delta]$ are occurring at the time point $i\Delta$. This is a renewal risk model

$$C_t^\Delta = u + ct - \sum_{i=1}^{N_t^\Delta} Y_i^\Delta \;,$$

where $\{N_t^\Delta\}$ is a renewal process with interarrival time distribution $F_\Delta(t) = \mathbb{1}_{\{t \ge \Delta\}}$ and claim sizes $\{Y_i^\Delta\}$ with a compound mixed Poisson distribution. More explicitly,

$$G_\Delta(x) = \mathbb{P}[Y_i^\Delta \le x] = \mathbb{P}\Bigl[\sum_{i=1}^{N_\Delta} Y_i \le x\Bigr] .$$

Given L_1 the conditional expected number of claims and thus the Poisson parameter is ΔL_1. From (1.2) we already know the moment generating function

$$M_{Y^\Delta}(r) = M_L(\Delta(M_Y(r) - 1))$$

of Y_i^Δ where $M_L(r)$ is the moment generating function of the intensity levels L_i.

In order to avoid $\psi(u) = 1$ we need that the random walk $\{C_{k\Delta}\}$ tends to infinity, or equivalently that $\mathbb{E}[C_\Delta - u] > 0$. The mean value of the claims up to time Δ is $\mathbb{E}[L_1]\mu\Delta$ [see (1.1)] and the income is $c\Delta$. Thus the net profit condition must be

$$c > \mathbb{E}[L_1]\mu \, .$$

We assume in the future that the net profit condition is fulfilled.

The ruin probabilities of $\{C_t\}$ and of $\{C_t^\Delta\}$ can be compared in the following way.

Lemma 7.1 *For $\psi(u)$ and $\psi_\Delta(u)$ as defined above we have*

$$\psi(u + c\Delta) \le \psi_\Delta(u) \le \psi(u) \le \psi_\Delta(u - c\Delta) \, .$$

Proof Assume that there exists a $t_0 > 0$ such that $C_{t_0} + c\Delta < 0$, i.e. ruin occurs for the Ammeter risk model with initial capital $u + c\Delta$. Let k_0 such that $t_0 \in ((k_0 - 1)\Delta, k_0\Delta]$. Then

$$0 > c\Delta + C_{t_0} = c\Delta + C_{k_0\Delta} - c(k_0\Delta - t_0) + \sum_{i=N_{t_0}+1}^{N_{k_0\Delta}} Y_i > C_{k_0\Delta} \, .$$

Thus $\psi(u + c\Delta) \le \psi_\Delta(u)$ by noting that $C_{k_0\Delta} = C_{k_0\Delta}^\Delta$. The last inequality follows from the first one and the second inequality is trivial. □

7.3 Lundberg's Inequality and the Cramér–Lundberg Approximation

7.3.1 The Ordinary Case

The first problem is to find the analogue of the Lundberg exponent for the Ammeter risk model. In the classical case and in the renewal case we constructed an exponential martingale. Here, as in the renewal case, the process $\{C_t\}$ is no longer a Markov process, but for each $k \in \mathbb{N}$ the process $\{C_{t+k\Delta} - C_{k\Delta}\}$ is independent of $\mathcal{F}_{k\Delta}$. Let us therefore only consider the time points $0, \Delta, 2\Delta, \ldots$.

Lemma 7.2 *Let $\{C_t\}$ be an Ammeter risk model. For any $r \in \mathbb{R}$ such that $M_L(\Delta(M_Y(r) - 1)) < \infty$ let $\theta(r)$ be the unique solution to*

$$e^{-(\theta(r)+cr)\Delta} M_L(\Delta(M_Y(r) - 1)) = 1 .$$

Then the discrete time process

$$\{\exp\{-rC_{k\Delta} - \theta(r)k\Delta\}\}$$

is a martingale. Moreover, $\theta(r)$ is strictly convex, $\theta(0) = 0$ and $\theta'(0) = \mathbb{E}[L_1]\mu - c < 0$.

Proof Considering the process at the time points $0, \Delta, 2\Delta, \ldots$ we find $C_{k\Delta} = C_{k\Delta}^{\Delta}$. These points are exactly the claim times of the renewal risk process $\{C_{k\Delta}^{\Delta}\}$. The assertion follows from Lemma 6.1 and from (6.3). □

Because $\theta(r)$ is a convex function and $\theta(0) = 0$ there may exist a second solution R to the equation $\theta(r) = 0$. R is, if it exists, unique and strictly positive. R is then called the *Lundberg exponent* or *adjustment coefficient* of the Ammeter risk model.

Next we prove Lundberg's inequality.

Theorem 7.3 *Let $\{C_t\}$ be an Ammeter risk model and assume that the Lundberg exponent R exists. Then*

$$\psi(u) < e^{cR\Delta} e^{-Ru} .$$

If there exists an $r > R$ such that $M_L(\Delta(M_Y(r) - 1)) < \infty$ then R is the correct exponent in the sense that there exists a constant $\underline{C} > 0$ such that

$$\psi(u) \geq \underline{C} e^{-Ru} .$$

Remark 7.4 The constant $e^{cR\Delta}$ might be too large, especially if Δ is large. If possible, one should therefore try to find an alternative for the constant. An upper bound, based on the martingale (7.2), is found in [55] and [110]. However, one has to accept that the constant might be larger than 1. Recall that this was also the case for Lundberg's inequality in the general case of the renewal risk model. ■

Proof Using Lemma 7.1 the proof becomes almost trivial. From Theorem 6.6 it follows that $\psi_\Delta(u) < \mathrm{e}^{-Ru}$. Thus

$$\psi(u) \le \psi_\Delta(u - c\Delta) < \mathrm{e}^{-R(u-c\Delta)} = \mathrm{e}^{cR\Delta}\mathrm{e}^{-Ru}\,.$$

If there exists an $r > R$ such that $M_L(\Delta(M_Y(r) - 1)) < \infty$ then by Theorem 6.8 there exists a constant $\underline{C} > 0$ such that $\psi_\Delta(u) \ge \underline{C}\mathrm{e}^{-Ru}$. Thus also

$$\psi(u) \ge \psi_\Delta(u) \ge \underline{C}\mathrm{e}^{-Ru}\,.$$

□

The constant $\mathrm{e}^{cR\Delta}$ seems to increase to infinity as Δ tends to infinity. However, R also depends on Δ. At least in some cases $\mathrm{e}^{cR\Delta}$ remains bounded as $\Delta \to \infty$.

Proposition 7.5 *Assume that R exists and that there exists a solution $r_0 \ne 0$ to*

$$M_L(\mu r)\mathrm{e}^{-cr} = 1\,.$$

Then

$$\mathrm{e}^{cR\Delta} \le \mathrm{e}^{cr_0}\,. \tag{7.1}$$

In particular, the condition will be fulfilled if R exists and if there exists an $\ell_\infty \le \infty$ such that $M_L(r) < \infty \iff r < \ell_\infty$.

Remark 7.6 Note that r_0 depends on the claim size distribution via μ only. ■

Proof The function $\log(M_L(\mu r)) - cr$ has the second derivative

$$\mu^2\left(\frac{M_L''(\mu r)}{M_L(\mu r)} - \left(\frac{M_L'(\mu r)}{M_L(\mu r)}\right)^2\right) > 0$$

(see Lemma 1.13) and is therefore strictly convex. Because

$$\mu\left(\frac{M_L'(0)}{M_L(0)}\right) - c = \mu\mathbb{E}[L_i] - c < 0$$

r_0 has to be strictly positive. In particular, if

$$M_L(\mu r)\mathrm{e}^{-cr} = \exp\{\log(M_L(\mu r)) - cr\} \le 1$$

then $r \le r_0$. If $M_L(r) < \infty \iff r < \ell_\infty$ then by monotone convergence

$$\lim_{r\uparrow\ell_\infty} M_L(r) = M_L(\ell_\infty) = \infty\,.$$

Thus by the convexity a strictly positive solution to $\log(M_L(\mu r)) - cr = 0$ has to exist.

Recall that $M_L(r)$ is increasing and that $M_Y(r) - 1 \ge rM'_Y(0) = r\mu$. Since

$$1 = \mathrm{e}^{-cR\Delta} M_L(\Delta(M_Y(R) - 1)) \ge \mathrm{e}^{-cR\Delta} M_L(\Delta R\mu)$$

it follows from the consideration above that $R\Delta \le r_0$. □

Example 7.7 Assume that $L_i \sim \Gamma(\gamma, \alpha)$. The case $\Delta = 1$ was the case considered in Ammeter's original paper [2]. In order to obtain the Lundberg exponent we have to solve the equation

$$\mathrm{e}^{-cr\Delta}\Bigl(\frac{\alpha}{\alpha - \Delta(M_Y(r) - 1)}\Bigr)^{\gamma} = 1$$

or equivalently

$$\mathrm{e}^{-cr\Delta/\gamma}\frac{\alpha}{\alpha - \Delta(M_Y(r) - 1)} = 1\,.$$

This can be written as

$$[\alpha - \Delta(M_Y(r) - 1)]\,\mathrm{e}^{cr\Delta/\gamma} = \alpha\,.$$

It follows immediately that the above equation only admits a solution in closed form for special cases. For example, if $Y_i = 1$ is deterministic then one has to solve

$$\Delta\mathrm{e}^{(1+c\Delta/\gamma)r} - (\alpha + \Delta)\mathrm{e}^{(c\Delta/\gamma)r} + \alpha = 0\,.$$

One can only hope for a solution in closed form if $c\Delta/\gamma \in \mathbb{Q}$. For instance, if $c\Delta/\gamma = 1$ then $R = \log(\alpha/\Delta) = \log(\alpha c/\gamma)$.

Let us now consider (7.1). r_0 is the strictly positive solution to

$$\Bigl(\frac{\alpha}{\alpha - r\mu}\Bigr)^{\gamma}\mathrm{e}^{-cr} = 1$$

or

$$(\alpha - r\mu)\mathrm{e}^{cr/\gamma} = \alpha\,.$$

The latter equation also has to be solved numerically. ■

From Theorem 7.3 we find that

$$\underline{C} \le \liminf_{u\to\infty} \psi(u)\mathrm{e}^{Ru} \le \limsup_{u\to\infty} \psi(u)\mathrm{e}^{Ru} \le \mathrm{e}^{cR\Delta}\,.$$

The question is: does there exist a constant C such that

$$\lim_{u\to\infty} \psi(u)\mathrm{e}^{Ru} = C\,?$$

That is, does the Cramér–Lundberg approximation hold? This question is answered in the following theorem.

Theorem 7.8 *Let $\{C_t\}$ be an Ammeter risk model. Assume that R exists and that there exists an $r > R$ such that $M_L(\Delta(M_Y(r) - 1)) < \infty$. Then there exists a constant $0 < C \le e^{cR\Delta}$ such that*

$$\lim_{u\to\infty} \psi(u)e^{Ru} = C\,.$$

Proof This is a special case of Theorem 2 of [114]. □

We next want to derive an alternative proof of Lundberg's inequality. For this reason we have to find an exponential martingale in continuous time. In applications it is usually rather easy to find a martingale if one is dealing with Markov processes. However, we have remarked before that $\{C_t\}$ is not a Markov process. The future of the process depends on the present level λ_t of the intensity. Let us therefore consider the process $\{(C_t, \lambda_t)\}$, which, however, is also not a homogeneous Markov process. The future of the process also depends on time because the future intensity level is known until the next change of the intensity. Instead of considering $\{(C_t, \lambda_t, t)\}$ we consider $\{(C_t, \lambda_t, V_t)\}$ where V_t is the time remaining till the next change of the intensity, i.e.

$$V_t = k\Delta - t, \qquad \text{if } (k-1)\Delta \le t < k\Delta.$$

Remarks 7.9

(i) The process $\{\lambda_t\}$ cannot be observed directly. The only information on λ_t is the number of claims which occurred in the interval $[t + V_t - \Delta, t]$. Thus, provided L_i is not deterministic, we can, in the best case, find the distribution of λ_t given $\mathcal{F}_t^C$, the information generated by $\{C_s : s \le t\}$ up to time t. For our purposes, however, it is enough to work with the filtration $\mathcal{F}_t$ generated by $\{(C_t, \lambda_t, V_t)\}$. The results will not depend on the chosen filtration.

(ii) As an alternative we could consider the time elapsed since the last change of the intensity $W_t = \Delta - V_t$. For the present choice of the Markovisation see also Sect. 8.3.

We find the following martingale.

Lemma 7.10 *Let r such that $M_L(\Delta(M_Y(r) - 1)) < \infty$. Then the process*

$$M_t^r = e^{-rC_t} e^{(\lambda_t(M_Y(r)-1)-cr-\theta(r))V_t} e^{-\theta(r)t} \tag{7.2}$$

is a martingale with respect to the natural filtration of $\{(C_t, \lambda_t, V_t)\}$.

Proof Let first $(k-1)\Delta \le s < t < k\Delta$. Then $\lambda_t = \lambda_s$ and

$$\begin{aligned}\mathbb{E}[M_t^r \mid \mathcal{F}_s] &= \mathbb{E}\big[e^{-rC_t} e^{(\lambda_t(M_Y(r)-1)-cr-\theta(r))V_t} e^{-\theta(r)t} \bigm| \mathcal{F}_s\big] \\ &= M_s^r \mathbb{E}\big[e^{-r(C_t-C_s)} \bigm| \mathcal{F}_s\big] e^{-(\lambda_t(M_Y(r)-1)-cr)(t-s)} \\ &= M_s^r e^{(\lambda_t(M_Y(r)-1)-cr)(t-s)} e^{-(\lambda_t(M_Y(r)-1)-cr)(t-s)} = M_s^r\,.\end{aligned}$$

Let now $(k-1)\Delta \le s < t = k\Delta$. Then, using the definition of $\theta(r)$,

$$\begin{aligned}\mathbb{E}[M_t^r \mid \mathcal{F}_s] = M_s^r \mathbb{E}\big[e^{-r(C_t-C_s)} \bigm| \mathcal{F}_s\big] &\mathbb{E}\big[e^{(L_{k+1}(M_Y(r)-1)-cr-\theta(r))\Delta}\big] \\ &\times e^{-(\lambda_s(M_Y(r)-1)-cr-\theta(r))(t-s)} e^{-\theta(r)(t-s)} \\ = M_s^r e^{(\lambda_s(M_Y(r)-1)-cr)(t-s)} &M_L(\Delta(M_Y(r)-1)) e^{-(cr+\theta(r))\Delta} \\ &\times e^{-(\lambda_s(M_Y(r)-1)-cr)(t-s)} = M_s^r\,.\end{aligned}$$

The assertion follows by iteration. □

From the martingale stopping theorem we find

$$e^{-Ru} = \mathbb{E}[M_0^R] = \mathbb{E}[M_{\tau\wedge t}^R] \ge \mathbb{E}[M_\tau^R; \tau \le t]\,.$$

By monotone convergence it follows that

$$e^{-Ru} \ge \mathbb{E}[M_\tau^R; \tau < \infty] = \mathbb{E}[M_\tau^R \mid \tau < \infty]\mathbb{P}[\tau < \infty]\,,$$

from which

$$\psi(u) \le \frac{e^{-Ru}}{\mathbb{E}[M_\tau^R \mid \tau < \infty]}$$

follows. Note that $C_\tau < 0$ on $\{\tau < \infty\}$. Thus

$$\begin{aligned}\mathbb{E}[M_\tau^R \mid \tau < \infty] &> \mathbb{E}[e^{(\lambda_\tau(M_Y(R)-1)-cR)V_\tau} \mid \tau < \infty] \\ &> \mathbb{E}[e^{-cRV_\tau} \mid \tau < \infty] > e^{-cR\Delta}\,.\end{aligned}$$

We have recovered Lundberg's inequality

$$\psi(u) < e^{cR\Delta} e^{-Ru}\,.$$

7.3.2 *The General Case*

The second proof of Lundberg's inequality is particularly useful if the process does not start at a point where the intensity changes. For instance, the premium for the next year has to be determined before the end of the year. In this case the actuary has some information about the present intensity via the number of claims which occurred in the first part of the year. Let us therefore assume that (λ_0, V_0) has the joint distribution function $F^1(\ell, v)$. We assume that $\mathbb{P}[V_0 \le \Delta] = 1$. By allowing V_0 to be stochastic we can also treat cases where no information on the intensity changing times is available. This seems a little strange from a practical point of view, but it is nevertheless interesting from a theoretical point of view. One special case is the case of a stationary intensity process where

$$F^1(\ell, v) = \Bigl(\frac{v}{\Delta}\mathbb{1}_{0\le v\le\Delta} + \mathbb{1}_{v>\Delta}\Bigr)H(\ell)\,.$$

We get the following version of Lundberg's inequality.

Theorem 7.11 *Let $\{C_t\}$ be an Ammeter risk model. Assume that the Lundberg exponent R exists and that*

$$\mathbb{E}[e^{\lambda_0(M_Y(R)-1)V_0}] < \infty\,.$$

Then

$$\psi(u) < e^{cR\Delta}\mathbb{E}[e^{(\lambda_0(M_Y(R)-1)-cR)V_0}]e^{-Ru}\,.$$

Remark 7.12 The condition

$$\mathbb{E}\bigl[e^{\lambda_0(M_Y(R)-1)V_0}\bigr] < \infty$$

will be fulfilled in all reasonable situations. For instance, if (λ_0, V_0) is distributed such that $\{\lambda_t\}$ becomes stationary, then

$$\begin{aligned}\mathbb{E}\bigl[e^{\lambda_0(M_Y(R)-1)V_0}\bigr] &= \frac{1}{\Delta}\int_0^\Delta M_L(v(M_Y(R)-1))\,dv\\ &< M_L(\Delta(M_Y(R)-1)) < \infty\,.\end{aligned}$$

■

Proof From the stopping theorem it follows that

$$e^{-Ru}\mathbb{E}\bigl[e^{(\lambda_0(M_Y(R)-1)-cR)V_0}\bigr] = \mathbb{E}[M_0^R] = \mathbb{E}[M_{\tau\wedge t}^R] \ge \mathbb{E}[M_\tau^R; \tau \le t]\,.$$

The assertion now follows as before. □

For the Cramér–Lundberg approximation we need the following

Lemma 7.13 *Assume that an $r > R$ exists such that $M_L(\Delta(M_Y(r)-1)) < \infty$ and that*

$$\mathbb{E}[e^{\lambda_0(M_Y(r)-1)V_0}] < \infty\,.$$

Then

$$\lim_{u\to\infty} \mathbb{P}[\tau \le V_0 \mid C_0 = u]e^{Ru} = 0\,.$$

Proof Again using the stopping theorem we get

$$e^{-ru}\mathbb{E}[e^{(\lambda_0(M_Y(r)-1)-cr-\theta(r))V_0}] = \mathbb{E}[M_0^r] = \mathbb{E}[M^r_{\tau\wedge V_0}] \ge \mathbb{E}[M^r_\tau;\, \tau \le V_0]\,.$$

As before it follows that

$$\begin{aligned}\mathbb{P}[\tau \le V_0] &\le e^{(rc+\theta(r))\Delta}\mathbb{E}[e^{(\lambda_0(M_Y(r)-1)-cr-\theta(r))V_0}]e^{-ru}\\ &< e^{(rc+\theta(r))\Delta}\mathbb{E}[e^{\lambda_0(M_Y(r)-1)V_0}]e^{-ru}\,,\end{aligned}$$

where we used that $\theta(r) > 0$. Multiplying by e^{Ru} and then letting $u \to \infty$ yields the assertion. □

The Cramér–Lundberg approximation in the general case can now be proved.

Theorem 7.14 *Let $\{C_t\}$ be an Ammeter risk model. Assume that R exists and that there exists an $r > R$ such that both $M_L(\Delta(M_Y(r) - 1)) < \infty$ and*

$$\mathbb{E}[e^{\lambda_0(M_Y(r)-1)V_0}] < \infty\,.$$

Then

$$\lim_{u\to\infty} \psi(u)e^{Ru} = \mathbb{E}[e^{(\lambda_0(M_Y(R)-1)-cR)V_0}]\,C\,,$$

where C is the constant obtained in Theorem 7.8.

Proof Let $\psi^o(u)$ denote the ruin probability in the ordinary case and define

$$B(x;\ell,v) := \mathbb{P}[C_v - u \le x \mid \lambda_0 = \ell, V_0 = v]\,.$$

Then

$$\begin{aligned}\psi(u) = \int_{\ell=0}^\infty \int_{v=0}^\Delta \int_{y=-u}^{cv} &\psi^o(u+y)\mathbb{P}[\tau > v \mid \lambda_0 = \ell, V_0 = v, C_v = u+y]\\ &\times B(dy;\ell,v)F^1(d\ell,dv) + \mathbb{P}[\tau \le V_0]\,.\end{aligned}$$

We multiply the above equation by e^{Ru}. The last term tends to 0 by Lemma 7.13. The first term on the right-hand side can be written as

$$\begin{aligned}\int_{\ell=0}^\infty \int_{v=0}^\Delta \int_{y=-u}^{cv} &\psi^o(u+y)e^{R(u+y)}\mathbb{P}[\tau > v \mid \lambda_0 = \ell, V_0 = v, C_v = u+y]\\ &\times e^{-Ry}B(dy;\ell,v)F^1(d\ell,dv)\,.\end{aligned}$$

Noting that by Theorem 7.8 $\psi^{\mathrm{o}}(u+y)\mathrm{e}^{R(u+y)}$ is uniformly bounded by a constant $\bar{C}$ we see that

$$\begin{aligned}\int_{\ell=0}^{\infty}\int_{v=0}^{\Delta}\int_{y=-u}^{cv} &\psi^{\mathrm{o}}(u+y)\mathrm{e}^{R(u+y)}\mathbb{P}[\tau>v \mid \lambda_0=\ell, V_0=v, C_v=u+y] \\ &\times \mathrm{e}^{-Ry}B(\mathrm{d}y;\ell,v)F^1(\mathrm{d}\ell,\mathrm{d}v) \\ \le&\bar{C}\mathbb{E}[\mathrm{e}^{-R(C_{V_0}-u)}]=\bar{C}\mathbb{E}[\mathrm{e}^{(\lambda_0(M_Y(R)-1)-cR)V_0}]<\infty\,.\end{aligned}$$

We can therefore interchange integration and limit. Because $\psi^{\mathrm{o}}(u+y)\mathrm{e}^{R(u+y)}$ tends to C and $\mathbb{P}[\tau>v \mid \lambda_0=\ell, V_0=v, C_v=u+y]$ tends to 1 as $u\to\infty$ the desired limit is

$$\begin{aligned}\lim_{u\to\infty}\psi(u)\mathrm{e}^{Ru}&=C\int_{\ell=0}^{\infty}\int_{v=0}^{\Delta}\int_{y=-\infty}^{cv}\mathrm{e}^{-Ry}B(\mathrm{d}y;\ell,v)F^1(\mathrm{d}\ell,\mathrm{d}v)\\ &=\mathbb{E}[\mathrm{e}^{(\lambda_0(M_Y(R)-1)-cR)V_0}]\,C\,.\end{aligned}$$

□

7.4 The Subexponential Case

A subexponential behaviour of the ruin probability in the Ammeter risk model can be caused by two sources. First the claim sizes can be subexponential. More explicitly, if there exists an $\varepsilon>0$ such that $M_L(\varepsilon)<\infty$ and Y_i has a subexponential distribution, then Y_i^{Δ} has a subexponential distribution. However, it is also possible that L_i has a distribution that makes Y_i^{Δ} subexponential. In any case we have the following

Proposition 7.15 *Assume that*

$$\frac{1}{\Delta\mathbb{E}[L_i]\mu}\int_0^x(1-G_{\Delta}(y))\,\mathrm{d}y$$

is subexponential. Then

$$\lim_{u\to\infty}\frac{\psi(u)}{\int_u^{\infty}(1-G_{\Delta}(y))\,\mathrm{d}y}=\frac{1}{c\Delta-\Delta\mathbb{E}[L_i]\mu}\,.$$

Proof From Theorem 6.14 the asymptotic behaviour in the assertion is true for $\psi(u)$ replaced by $\psi_{\Delta}(u)$. It is therefore sufficient to show that $\psi(u)/\psi_{\Delta}(u)$ tends to 1 as $u\to\infty$. We have from Lemma 7.1

$$1\le\frac{\psi(u)}{\psi_{\Delta}(u)}\le\frac{\psi_{\Delta}(u-c\Delta)}{\psi_{\Delta}(u)}\,.$$

Using Theorem 6.14 and Lemma F.2 we obtain

$$\lim_{u\to\infty} \frac{\psi_\Delta(u - c\Delta)}{\psi_\Delta(u)} = \lim_{u\to\infty} \frac{\int_{u-c\Delta}^\infty (1 - G_\Delta(y))\,\mathrm{d}y}{\int_u^\infty (1 - G_\Delta(y))\,\mathrm{d}y} = 1\,.$$

This proves the assertion. □

The next result deals with the tail of compound distributions with a regularly varying tail.

Lemma 7.16 *Let N be a random variable with values in $\mathbb{N}$ such that $\mathbb{E}[N] < \infty$, $\{Y_i\}$ be a sequence of positive iid. random variables independent of N with distribution function $G(y)$ and mean μ, and let $\tilde{G}(y)$ be the distribution function of $\sum_{i=1}^N Y_i$. Let $\delta > 1$ and $L(x)$ be a slowly varying function. Assume that*

$$\lim_{x\to\infty} \frac{x^\delta(1 - G(x))}{L(x)} = \varphi, \qquad \lim_{n\to\infty} \frac{n^\delta \mathbb{P}[N > n]}{L(n)} = \gamma$$

where φ and γ are positive constants. Then

$$\lim_{x\to\infty} \frac{x^\delta(1 - \tilde{G}(x))}{L(x)} = \mu^\delta\gamma + \mathbb{E}[N]\varphi\,.$$

Proof See [124]. □

Next we link the distribution function $H(\ell)$ of L_i to the distribution function of N_Δ.

Lemma 7.17 *Assume that there exists a $\delta > 0$ and a slowly varying function $L(x)$ such that*

$$\lim_{\ell\to\infty} \frac{(1 - H(\ell))\ell^\delta}{L(\ell)} = 0\,.$$

Then

$$\lim_{n\to\infty} \frac{n^\delta \mathbb{P}[N_\Delta > n]}{L(n)} = 0\,.$$

Proof First note that for $n \ge 1$

$$\begin{aligned}
\mathbb{P}[N_\Delta > n] &= \int_0^\infty \sum_{m=n+1}^\infty \frac{(\Delta\ell)^m}{m!} \mathrm{e}^{-\Delta\ell}\,\mathrm{d}H(\ell) \\
&= \Delta \int_0^\infty \sum_{m=n+1}^\infty \int_0^\ell \left(\frac{(\Delta y)^{m-1}}{(m-1)!} - \frac{(\Delta y)^m}{m!}\right) \mathrm{e}^{-\Delta y}\,\mathrm{d}y\,\mathrm{d}H(\ell) \\
&= \Delta \int_0^\infty \int_y^\infty \sum_{m=n+1}^\infty \left(\frac{(\Delta y)^{m-1}}{(m-1)!} - \frac{(\Delta y)^m}{m!}\right) \mathrm{e}^{-\Delta y}\,\mathrm{d}H(\ell)\,\mathrm{d}y
\end{aligned}$$

$$= \Delta \int_0^\infty \frac{(\Delta y)^n}{n!} e^{-\Delta y}(1 - H(y))\, dy$$
$$= \int_0^\infty \frac{y^n}{n!} e^{-y}(1 - H(y/\Delta))\, dy\,.$$

Choose now $a \le e^{-1}$. Then

$$\int_0^{an} \frac{y^n}{n!} e^{-y}(1 - H(y/\Delta))\, dy \le \int_0^{an} \frac{y^n}{n!} e^{-y}\, dy \le \frac{(an)^{n+1}}{n!} e^{-an}$$
$$\le \frac{n^{n+1}}{n!} e^{-an-n}$$

because $y^n e^{-y}$ is increasing in the interval $(0, n)$. By Stirling's formula there exists a $K > 0$ such that $n! \ge K n^{n+\frac{1}{2}} e^{-n}$, from which it follows that

$$n^\delta L(n)^{-1} \int_0^{an} \frac{y^n}{n!} e^{-y}(1 - H(y/\Delta))\, dy$$

converges to 0 as $n \to \infty$. The latter follows because $x^{-\varepsilon}/L(x)$ converges to 0 for any $\varepsilon > 0$.

For the remaining part we obtain

$$\int_{an}^\infty \frac{y^n}{n!} e^{-y}(1 - H(y/\Delta))\, dy \le \int_{an}^\infty \frac{1}{\Gamma(n+1)} y^n e^{-y}\, dy\, (1 - H(an/\Delta))$$
$$\le 1 - H(an/\Delta)\,.$$

Thus

$$n^\delta L(n)^{-1} \int_{an}^\infty \frac{y^n}{n!} e^{-y}(1 - H(y/\Delta))\, dy$$

converges to 0 because

$$n^\delta L(n)^{-1}(1 - H(an/\Delta)) = \left(\frac{\Delta}{a}\right)^\delta \frac{(an/\Delta)^\delta (1 - H(an/\Delta))}{L(an/\Delta)} \frac{L(an/\Delta)}{L(n)}$$

converges to 0. □

Lemma 7.18 *Let $L(x)$ be a locally bounded slowly varying function and let $\delta > 0$. Then*

$$\lim_{y\to\infty} \frac{\int_0^\infty x^{y-\delta} e^{-x} L(x)\, dx}{\Gamma(y - \delta + 1) L(y)} = 1\,.$$

Proof There exists a constant K such that $L(x) \le K x^\delta$. Assume that y is so large that $L(y) > y^{-\delta}$. Then

$$\int_0^{e^{-1}y} \frac{x^{y-\delta} e^{-x} L(x)}{\Gamma(y-\delta+1)} \, dx \Big/ L(y) \le \frac{K y^\delta}{\Gamma(y-\delta+1)} \int_0^{e^{-1}y} x^y e^{-x} \, dx$$
$$< \frac{K y^{y+\delta+1} e^{-(1+e^{-1})y}}{\Gamma(y-\delta+1)} ,$$

which converges to 0 as $y \to \infty$ by Stirling's formula. The last inequality follows because $x^y e^{-x}$ is increasing on $[0, y]$.

Analogously

$$\int_{2y}^{\infty} \frac{x^{y-\delta} e^{-x} L(x)}{\Gamma(y-\delta+1)} \, dx \Big/ L(y) \le \frac{K y^\delta}{\Gamma(y-\delta+1)} \int_{2y}^{\infty} x^y e^{-x/2} e^{-x/2} \, dx$$
$$\le \frac{K y^\delta (2y)^y}{\Gamma(y-\delta+1)} e^{-y} 2 e^{-y}$$

which also converges to 0 as $y \to \infty$ by Stirling's formula.

Because $L(xy)/L(y)$ converges uniformly as $y \to \infty$ for $e^{-1} \le x \le 2$ it follows that

$$\lim_{y\to\infty} \frac{\int_{e^{-1}y}^{2y} x^{y-\delta} e^{-x} L(x) \, dx}{\Gamma(y-\delta+1) L(y)} = \lim_{y\to\infty} \frac{\int_{e^{-1}y}^{2y} x^{y-\delta} e^{-x} \, dx}{\Gamma(y-\delta+1)} = 1 ,$$

where the last equality follows from the considerations above with $L(x) = 1$. □

Lemma 7.19 *Assume that there exists a $\delta > 0$ and a locally bounded slowly varying function $L(x)$ such that*

$$\lim_{\ell\to\infty} \frac{(1-H(\ell))\ell^\delta}{L(\ell)} = 1 .$$

Then

$$\lim_{n\to\infty} \frac{n^\delta \mathbb{P}[N_\Delta > n]}{L(n)} = \Delta^\delta .$$

Proof Let us consider

$$\left| \frac{\mathbb{P}[N_\Delta > n] - \Delta^\delta \int_0^\infty \frac{\ell^{n-\delta}}{n!} e^{-\ell} L(\ell/\Delta) \, d\ell}{\Delta^\delta \int_0^\infty \frac{\ell^{n-\delta}}{n!} e^{-\ell} L(\ell/\Delta) \, d\ell} \right|$$
$$= \left| \frac{\int_0^\infty \ell^{n-\delta} e^{-\ell} L(\ell/\Delta) \left(\frac{(\ell/\Delta)^\delta (1-H(\ell/\Delta))}{L(\ell/\Delta)} - 1 \right) d\ell}{\int_0^\infty \ell^{n-\delta} e^{-\ell} L(\ell/\Delta) \, d\ell} \right| .$$

Choose x_0 such that

$$1 - \frac{\varepsilon}{2} < \frac{(\ell/\Delta)^\delta (1-H(\ell/\Delta))}{L(\ell/\Delta)} < 1 + \frac{\varepsilon}{2}$$

for $\ell \ge x_0$. Then

$$\left|\frac{\int_{x_0}^\infty \ell^{n-\delta} e^{-\ell} L(\ell/\Delta)\left(\frac{(\ell/\Delta)^\delta(1-H(\ell/\Delta))}{L(\ell/\Delta)} - 1\right) d\ell}{\int_0^\infty \ell^{n-\delta} e^{-\ell} L(\ell/\Delta)\, d\ell}\right| < \frac{\varepsilon}{2}\,.$$

The remaining term can be estimated, where $\bar{L} = \max\{L(x/\Delta) : x \le x_0\}$,

$$\frac{\Delta^{-\delta}\int_0^{x_0} \ell^n e^{-\ell}(1-H(\ell/\Delta))\, d\ell + \int_0^{x_0} \ell^{n-\delta} e^{-\ell} L(\ell/\Delta)\, d\ell}{\int_0^\infty \ell^{n-\delta} e^{-\ell} L(\ell/\Delta)\, d\ell}$$
$$\le \frac{\Delta^{-\delta} x_0^{n+1} + x_0^{n-\delta+1}\bar{L}}{\int_{2x_0}^\infty \ell^{n-\delta} e^{-\ell} L(\ell/\Delta)\, d\ell} \le \left(\frac{1}{2}\right)^{n-\delta} \frac{\Delta^{-\delta} x_0^{\delta+1} + x_0\bar{L}}{\int_{2x_0}^\infty e^{-\ell} L(\ell/\Delta)\, d\ell}\,.$$

Choosing n large enough the latter can be made smaller than $\varepsilon/2$. Thus

$$\lim_{n\to\infty} \frac{\mathbb{P}[N_\Delta > n]}{\int_0^\infty \frac{\ell^{n-\delta}}{n!} e^{-\ell} L(\ell/\Delta)\, d\ell} = \Delta^\delta\,.$$

It follows from Stirling's formula that

$$\lim_{n\to\infty} \frac{n^\delta \Gamma(n-\delta+1)}{n!} = 1$$

and the assertion follows from Lemma 7.18 noting that $L(x/\Delta)$ is also a slowly varying function. □

We can now prove the asymptotic behaviour in the case where the heavier tail of G and H is regularly varying.

Theorem 7.20 *Assume that there exist a slowly varying function $L(x)$ and positive constants δ, ω and φ with $\delta > 1$ and $\omega + \varphi > 0$ such that*

$$\lim_{\ell\to\infty} \frac{\ell^\delta(1-H(\ell))}{L(\ell)} = \omega\,, \qquad \lim_{x\to\infty} \frac{x^\delta(1-G(x))}{L(x)} = \varphi\,.$$

Then

$$\lim_{u\to\infty} \frac{\psi(u)u^{\delta-1}}{L(u)} = \frac{\varphi\mathbb{E}[L_i]\Delta + \omega(\mu\Delta)^\delta}{(c-\mathbb{E}[L_i]\mu)\Delta(\delta-1)}\,.$$

Proof By Lemmas 7.17 and 7.19

$$\lim_{n\to\infty} \frac{n^\delta \mathbb{P}[N_\Delta > n]}{L(n)} = \omega\Delta^\delta\,.$$

From Lemma 7.16 one obtains

$$\lim_{x\to\infty} \frac{x^\delta(1 - G_\Delta(x))}{L(x)} = \mathbb{E}[L_i]\Delta\varphi + \mu^\delta \omega \Delta^\delta .$$

The result now follows from Proposition 7.15 and Karamata's theorem. □

Example 7.21 Assume again that $L_i \sim \Gamma(\gamma, \alpha)$. Assume that the claims are Pa(δ, β) distributed with $\delta > 1$. Then

$$\lim_{x\to\infty} x^\delta(1 - G(x)) = \beta^\delta$$

and

$$\lim_{x\to\infty} x^\delta(1 - H(x)) = \lim_{x\to\infty} \frac{\alpha^\gamma x^{\gamma-1} e^{-\alpha x}}{\Gamma(\gamma)\delta x^{-(\delta+1)}} = 0 .$$

Thus

$$\lim_{u\to\infty} u^{\delta-1}\psi(u) = \frac{\beta^\delta \gamma/\alpha\, \Delta}{(c - \gamma/\alpha\, \beta/(\delta-1))\Delta(\delta-1)} = \frac{\beta^\delta \gamma}{(c\alpha(\delta-1) - \gamma\beta)} .$$

■

Example 7.22 Let L_i be Pa(δ, β) distributed and assume that

$$\lim_{x\to\infty} x^\delta(1 - G(x)) = 0 .$$

Then it follows that

$$\lim_{u\to\infty} u^{\delta-1}\psi(u) = \frac{\beta^\delta (\mu\Delta)^\delta}{(c - \beta/(\delta-1)\, \mu)\Delta(\delta-1)} = \frac{(\beta\mu\Delta)^\delta}{(c(\delta-1) - \beta\mu)\Delta} .$$

■

7.5 Finite Time Lundberg Inequalities

Let us now return to the small claim case. As in the classical case it is possible to find exponential inequalities for ruin probabilities of the form

$$\mathbb{P}[\underline{y}u < \tau \le \overline{y}u] .$$

Recall that

$$M_t^r = e^{-rC_t} e^{(\lambda_t(M_Y(r)-1)-cr-\theta(r))V_t} e^{-\theta(r)t}$$

is a martingale provided $M_L(\Delta(M_Y(r) - 1)) < \infty$.

Let $0 \le \underline{y} < \overline{y} < \infty$. Using the stopping theorem we get

$$\begin{aligned} e^{-ru}\mathbb{E}[e^{(\lambda_0(M_Y(r)-1)-cr-\theta(r))V_0}] &= \mathbb{E}[M_0^r] = \mathbb{E}[M_{\tau\wedge\overline{y}u}] \\ &> \mathbb{E}[e^{-rC_\tau}e^{(\lambda_\tau(M_Y(r)-1)-cr-\theta(r))V_\tau}e^{-\theta(r)\tau} \mid \underline{y}u < \tau \le \overline{y}u]\mathbb{P}[\underline{y}u < \tau \le \overline{y}u] \\ &> e^{-(cr+\theta(r))\Delta}e^{-\max\{\theta(r)\underline{y},\theta(r)\overline{y}\}u}\mathbb{P}[\underline{y}u < \tau \le \overline{y}u]\,. \end{aligned}$$

Thus

$$\begin{aligned} &\mathbb{P}[\underline{y}u < \tau \le \overline{y}u] \\ &\qquad < e^{(cr+\theta(r))\Delta}\mathbb{E}[e^{(\lambda_0(M_Y(r)-1)-cr-\theta(r))V_0}]e^{-\min\{r-\theta(r)\underline{y},r-\theta(r)\overline{y}\}u} \end{aligned}$$

and we can define the finite time Lundberg exponent.

$$R(\underline{y},\overline{y}) := \sup\{\min\{r-\theta(r)\underline{y}, r-\theta(r)\overline{y}\} : r \in \mathbb{R}\}\,.$$

Let $r_\infty = \sup\{r \ge 0 : M_L(\Delta(M_Y(r)-1)) < \infty\}$ and assume that $r_\infty > R$. We have discussed $R(\underline{y},\overline{y})$ in the classical case (Sect. 5.14). Let r_y be the argument maximising $r - \theta(r)y$. We obtain the critical value

$$y_0 = (\theta'(R))^{-1} = \Bigl(\frac{M_L'(\Delta(M_Y(R)-1))M_Y'(R)}{M_L(\Delta(M_Y(R)-1))} - c\Bigr)^{-1}$$

and the two finite time Lundberg inequalities

$$\begin{aligned} &\mathbb{P}[0 < \tau \le yu \mid C_0 = u] \\ &\qquad < e^{(cr_y+\theta(r_y))\Delta}\mathbb{E}[e^{(\lambda_0(M_Y(r_y)-1)-cr_y-\theta(r_y))V_0}]e^{-R(0,y)u}\,, \end{aligned}$$

where $R(0,y) > R$ if $y < y_0$, and

$$\begin{aligned} &\mathbb{P}[yu < \tau < \infty \mid C_0 = u] \\ &\qquad < e^{(cr_y+\theta(r_y))\Delta}\mathbb{E}[e^{(\lambda_0(M_Y(r_y)-1)-cr_y-\theta(r_y))V_0}]e^{-R(y,\infty)u}\,, \end{aligned}$$

where $R(y,\infty) > R$ if $y > y_0$. The next result can be proved in the same way as Theorem 5.38.

Theorem 7.23 *Assume that $R < r_\infty$. Then*

$$\frac{\tau}{u} \longrightarrow y_0$$

in probability on the set $\{\tau < \infty\}$. □

Bibliographical Remarks

For a comprehensive introduction to mixed Poisson processes, see [77]. The Ammeter risk process was introduced by Ammeter [2]. The process, as defined here, was first considered by Björk and Grandell [22], where Lundberg's inequality was also proved. In the latter publication, the interval for which the intensity is constant is stochastic. The approach used here can be found in [76]. In particular, Lemma 7.1, Propositions 7.5, 7.15, Lemma 7.17 and Theorem 7.20 are taken from [76]. A more general discussion of the subexponential case is found in [14]. Lemma 7.10 was first obtained in [55]. Lemma 7.19 goes back to [77]. The finite time Lundberg inequalities were first proved in [55]. Further literature on the topic can be found, for example, in [74, 110].

Chapter 8
Change of Measure Techniques

In the risk models considered so far we constructed exponential martingales. Using the stopping theorem we were able to prove an upper bound for the ruin probabilities. However, the technique did not allow us to prove a lower bound for the Cramér–Lundberg approximation. In the Ammeter risk model it had been possible to find a lower bound by means of a comparison with a renewal risk model. The reason that this was possible was that the lengths of the intervals, in which the intensity was constant, was deterministic. This made the interarrival times and the claim amounts in the corresponding renewal risk model independent.

In order to be able to treat models where the intervals with constant intensity are stochastic we have to find another method. One possibility is the following. Let $\{M_t\}$ be a strictly positive martingale. Suppose that $M_t \to 0$ on $\{\tau = \infty\}$ and that M_t is bounded on $\{\tau > t\}$. Then

$$M_0 = \mathbb{E}[M_{\tau \wedge t}] = \mathbb{E}[M_\tau; \tau \le t] + \mathbb{E}[M_t; \tau > t].$$

Now monotone convergence yields $\mathbb{E}[M_\tau; \tau \le t] \to \mathbb{E}[M_\tau; \tau < \infty]$ and bounded convergence yields $\mathbb{E}[M_t; \tau > t] \to 0$ as $t \to \infty$. Thus

$$M_0 = \mathbb{E}[M_\tau; \tau < \infty] = \mathbb{E}[M_\tau \mid \tau < \infty]\mathbb{P}[\tau < \infty].$$

This gives the formula

$$\mathbb{P}[\tau < \infty] = \frac{M_0}{\mathbb{E}[M_\tau \mid \tau < \infty]}.$$

The problem is that, in order to prove a Cramér–Lundberg approximation, one needs M_τ and $\{\tau < \infty\}$ to be independent. This is usually not an easy task to verify.

The methods we will use here are "change of measure techniques". We will see that exponential martingales can be used to change the measure. Under the new

H. Schmidli, *Risk Theory*, Springer Actuarial,
DOI:10.1007/978-3-319-72005-0_8

measure the risk model will be of the same type, except that the net profit condition will no longer be fulfilled and ruin will always occur within finite time. The quantity to calculate will depend on the capital at ruin only, and not on the time to ruin. Thus the ruin problem will become easier to handle, and the (possible) dependence between τ and C_τ will no longer matter.

We found the asymptotic behaviour of the ruin probability, but in both the renewal and the Ammeter case we were not able to give an explicit formula for the constant involved, and the results are only valid for ruin in infinite time. If we are interested in an exact figure for the (finite or infinite time) ruin probability or for the distribution of the severity of ruin then the only method available is simulation. However, the event of interest is rare, so we will have to simulate a lot of paths of the process which will never lead to ruin. If we are dealing with the event of ruin in infinite time, then we will have to decide at which point we declare the path as "not leading to ruin" because we have to stop the simulation within finite time.

The change of measure technique will give us the opportunity to express the quantity of interest as a quantity under the new measure, but because under the new measure ruin occurs almost surely, the problems discussed above vanish. Thus the change of measure technique provides a clever method of simulation. We can in fact show that it is in some sense the asymptotic optimal way to simulate a rare event, see [11].

8.1 The Method

Let $\{L_t\}$ be a martingale with $L_t \ge 0$ and $\mathbb{E}[L_0] = 1$. On $\mathcal{F}_t$ we define a measure with the Radon–Nikodym derivative L_t. That is, for $A \in \mathcal{F}_t$, we define

$$Q_t[A] := \mathbb{E}_{\mathbb{P}}[L_t; A] := \mathbb{E}_{\mathbb{P}}[L_t \mathbb{I}_A].$$

That Q_t is a probability measure is easy to verify.

In fact, it looks as though the measure also depends on t, but the next lemma shows that this is not the case.

Lemma 8.1

(i) *Let $t > s$ and $A \in \mathcal{F}_s \subset \mathcal{F}_t$. Then $Q_t[A] = Q_s[A]$.*

(ii) *Suppose it is possible to extend the measures Q_t to a measure Q on the whole σ-algebra $\mathcal{F}$. If T is a stopping time and $A \in \mathcal{F}_T$, such that $A \subset \{T < \infty\}$, then*

$$Q[A] := \mathbb{E}_{\mathbb{P}}[L_T; A].$$

Proof (i) This is an easy consequence of the martingale property

$$\mathbb{E}_{\mathbb{P}}[L_t; A] = \mathbb{E}_{\mathbb{P}}[\mathbb{E}_{\mathbb{P}}[L_t; A \mid \mathcal{F}_s]] = \mathbb{E}_{\mathbb{P}}[\mathbb{E}_{\mathbb{P}}[L_t \mid \mathcal{F}_s]; A] = \mathbb{E}_{\mathbb{P}}[L_s; A].$$

(ii) We have by the stopping theorem and the monotone convergence theorem

$$\begin{aligned}Q[A] &= \lim_{t\to\infty} Q[A\cap\{T\le t\}] = \lim_{t\to\infty} \mathbb{E}_{\mathbb{P}}[L_t;\, A\cap\{T\le t\}]\\ &= \lim_{t\to\infty} \mathbb{E}_{\mathbb{P}}[\mathbb{E}_{\mathbb{P}}[L_t \mid \mathcal{F}_T];\, A\cap\{T\le t\}] = \lim_{t\to\infty} \mathbb{E}_{\mathbb{P}}[L_{T\wedge t};\, A\cap\{T\le t\}]\\ &= \lim_{t\to\infty} \mathbb{E}_{\mathbb{P}}[L_T;\, A\cap\{T\le t\}] = \mathbb{E}_{\mathbb{P}}[L_T;\, A].\end{aligned}$$

□

One often needs conditional probabilities.

Lemma 8.2 *Let T be a stopping time, Y an $\mathcal{F}_T$-measurable and integrable random variable with $Y\mathbb{1}_{T<\infty} = Y$ and $\mathcal{G}\subset\mathcal{F}_T$ a σ-Algebra. We further assume that T is bounded, or that the measure Q can be extended to $\mathcal{F}$. Then*

$$\mathbb{E}_Q[Y \mid \mathcal{G}] = \frac{\mathbb{E}_{\mathbb{P}}[L_T Y \mid \mathcal{G}]}{\mathbb{E}_{\mathbb{P}}[L_T \mid \mathcal{G}]}.$$

In particular, for $s > t$ and an $\mathcal{F}_s$-measurable Y,

$$\mathbb{E}_Q[Y \mid \mathcal{F}_t] = \frac{\mathbb{E}_{\mathbb{P}}[L_s Y \mid \mathcal{F}_t]}{L_t}.$$

Proof Let Z be a bounded $\mathcal{G}$-measurable random variable. Then

$$\begin{aligned}\mathbb{E}_Q[YZ] &= \mathbb{E}_{\mathbb{P}}[L_T YZ] = \mathbb{E}_{\mathbb{P}}[\mathbb{E}_{\mathbb{P}}[L_T YZ \mid \mathcal{G}]] = \mathbb{E}_{\mathbb{P}}[\mathbb{E}_{\mathbb{P}}[L_T Y \mid \mathcal{G}]Z]\\ &= \mathbb{E}_{\mathbb{P}}\Bigl[\frac{\mathbb{E}_{\mathbb{P}}[L_T Y \mid \mathcal{G}]}{\mathbb{E}_{\mathbb{P}}[L_T \mid \mathcal{G}]}\mathbb{E}_{\mathbb{P}}[L_T \mid \mathcal{G}]Z\Bigr]\\ &= \mathbb{E}_{\mathbb{P}}\Bigl[\frac{\mathbb{E}_{\mathbb{P}}[L_T Y \mid \mathcal{G}]}{\mathbb{E}_{\mathbb{P}}[L_T \mid \mathcal{G}]}\mathbb{E}_{\mathbb{P}}[L_T Z \mid \mathcal{G}]\Bigr]\\ &= \mathbb{E}_{\mathbb{P}}\Bigl[\mathbb{E}_{\mathbb{P}}\Bigl[\frac{\mathbb{E}_{\mathbb{P}}[L_T Y \mid \mathcal{G}]}{\mathbb{E}_{\mathbb{P}}[L_T \mid \mathcal{G}]}L_T Z \Bigm| \mathcal{G}\Bigr]\Bigr]\\ &= \mathbb{E}_{\mathbb{P}}\Bigl[\frac{\mathbb{E}_{\mathbb{P}}[L_T Y \mid \mathcal{G}]}{\mathbb{E}_{\mathbb{P}}[L_T \mid \mathcal{G}]}L_T Z\Bigr] = \mathbb{E}_Q\Bigl[\frac{\mathbb{E}_{\mathbb{P}}[L_T Y \mid \mathcal{G}]}{\mathbb{E}_{\mathbb{P}}[L_T \mid \mathcal{G}]}Z\Bigr].\end{aligned}$$

This proves the first assertion. The second assertion follows for $\mathcal{G} = \mathcal{F}_t$ and $T = s$ since $\mathbb{E}_{\mathbb{P}}[L_s \mid \mathcal{F}_t] = L_t$. □

If $L_t > 0$ for all t, then $1/L_t$ is a Q-martingale with $\mathbb{E}_Q[1/L_0] = 1$. Indeed, $\mathbb{E}_Q[1/L_s \mid \mathcal{F}_t] = L_t^{-1}\mathbb{E}_{\mathbb{P}}[L_s/L_s \mid \mathcal{F}_t] = 1/L_t$ and $\mathbb{E}_Q[1/L_0] = \mathbb{E}_{\mathbb{P}}[L_0/L_0] = 1$. Using $1/L_t$ as the Radon–Nikodym derivative recovers the measure $\mathbb{P}$, since $\mathbb{E}_Q[1/L_t;\, A] = \mathbb{E}_{\mathbb{P}}[L_t/L_t;\, A] = \mathbb{P}[A]$ for $A\in\mathcal{F}_t$.

8.2 The Cramér–Lundberg Case

The simplest model we can use to illustrate the change of measure technique is the classical risk model. Let $\{C_t\}$ be a Cramér–Lundberg process and assume that the Lundberg exponent R exists and that there is an $r > R$ such that $M_Y(r) < \infty$. We have seen that for any r such that $M_Y(r) < \infty$ the process

$$L_t^r := \mathrm{e}^{-r(C_t-u)}\mathrm{e}^{-\theta(r)t}$$

is a martingale, where $\theta(r) = \lambda(M_Y(r) - 1) - cr$. This martingale has the desired properties. For $A \in \mathcal{F}_t$ we let

$$Q_r[A] := \mathbb{E}_{\mathbb{P}}[L_t^r; A].$$

We suppose here that $\{\mathcal{F}_t\}$ is the smallest right continuous filtration such that $\{C_t\}$ is adapted.

From the lemma below we can see that it is possible to extend the measures Q_r on $\mathcal{F}_t$ to the whole σ-algebra $\mathcal{F}$.

Lemma 8.3

(i) *$\{C_t\}$ is a Cramér–Lundberg process under the measure Q_r with claim size distribution $\tilde{G}(x) = \int_0^x \mathrm{e}^{ry}\,\mathrm{d}G(y)/M_Y(r)$ and claim arrival intensity $\tilde{\lambda} = M_Y(r)\lambda$. In particular, $\mathbb{E}_{Q_r}[C_1 - u] = -\theta'(r)$.*

(ii) *If $r \neq 0$ then Q_r and $\mathbb{P}$ are singular on the whole σ-algebra $\mathcal{F}$.*

Proof (i) For all $n \geq 1$ we have to find the joint distribution of the variables $T_1, T_2 - T_1, \dots, T_n - T_{n-1}$ and $Y_1, Y_2, \dots, Y_n$. Note that these random variables are independent under $\mathbb{P}$. Let $t \geq 0$, A be a Borel set of $\mathbb{R}_+^n$ and $B_1, B_2, \dots, B_n$ be Borel sets on $\mathbb{R}_+$. Then

$$\begin{aligned}
&Q_r[N_t = n, (T_1, T_2, \dots, T_n) \in A, Y_1 \in B_1, Y_2 \in B_2, \dots, Y_n \in B_n] \\
&\quad= \mathbb{E}_{\mathbb{P}}[L_t^r; N_t = n, (T_1, T_2, \dots, T_n) \in A, Y_1 \in B_1, Y_2 \in B_2, \dots, Y_n \in B_n] \\
&\quad= \mathbb{E}_{\mathbb{P}}\Bigl[\exp\Bigl\{r\sum_{i=1}^n Y_i\Bigr\}\exp\{-(\theta(r)+cr)t\}; N_t = n, (T_1, \dots, T_n) \in A, \\
&\qquad\qquad\qquad\qquad\qquad Y_1 \in B_1, Y_2 \in B_2, \dots, Y_n \in B_n\Bigr] \\
&\quad= \frac{(\lambda t)^n}{n!}\mathrm{e}^{-\lambda t-(\theta(r)+cr)t}\,\mathbb{P}[(T_1, T_2, \dots, T_n) \in A \mid N_t = n] \\
&\qquad\times \prod_{i=1}^n \mathbb{E}_{\mathbb{P}}[\mathrm{e}^{rY_i}; Y_i \in B_i]
\end{aligned}$$

$$= \frac{(\lambda M_Y(r)t)^n}{n!} e^{-\lambda M_Y(r)t} \mathbb{P}[(T_1, T_2, \ldots, T_n) \in A \mid N_t = n] \times \prod_{i=1}^n \frac{\mathbb{E}_{\mathbb{P}}[e^{rY_i}; Y_i \in B_i]}{M_Y(r)}.$$

Thus the number of claims is Poisson distributed with parameter $\lambda M_Y(r)$. Given $\{N_t\}$, the claim occurrence times do not change the conditional law. By Proposition C.3 the process $\{N_t\}$ is a Poisson process with rate $\lambda M_Y(r)$. The claim sizes are independent and have the claimed distribution. In particular

$$\mathbb{E}_{Q_r}[Y_i^n] = \frac{\int_0^\infty y^n e^{ry} \, dG(y)}{M_Y(r)} = \frac{M_Y^{(n)}(r)}{M_Y(r)},$$

from which (i) follows. Note that the measure can be extended to $\mathcal{F}$.

(ii) This follows because

$$Q_r\Big[\lim_{t\to\infty} \frac{C_t}{t} = -\theta'(r)\Big] = 1$$

and because $\theta'(r)$ is strictly increasing.

□

We next express the ruin probability using the measure Q_r,

$$\psi(u) = \mathbb{P}[\tau < \infty] = \mathbb{E}_{Q_r}[(L_\tau^r)^{-1}; \tau < \infty] = \mathbb{E}_{Q_r}[e^{rC_\tau} e^{\theta(r)\tau}; \tau < \infty] e^{-ru}.$$

The latter is much more complicated because the joint distribution of τ and C_τ has to be known unless $\theta(r) = 0$. Let us therefore choose $r = R$ and $Q = Q_R$. First observe that $\mathbb{E}_Q[C_1 - u] = -\theta'(R) < 0$. Thus the net profit condition is not fulfilled and $\tau < \infty$ a.s. under the measure Q. Then

$$\psi(u) = \mathbb{E}_Q[e^{RC_\tau}; \tau < \infty] e^{-Ru} = \mathbb{E}_Q[e^{RC_\tau}] e^{-Ru}.$$

Lundberg's inequality (Theorem 5.5) now becomes trivial. Because $C_\tau < 0$ one has $\psi(u) < e^{-Ru}$.

In order to prove the Cramér–Lundberg approximation we have to show that $\mathbb{E}_Q[e^{RC_\tau}]$ converges as $u \to \infty$. Denote by $B(x) = Q[C_\tau \ge -x \mid C_0 = 0]$ the descending ladder height distribution under Q. Note that (5.9) does not apply because the limit of $\psi_x(u)$ is no longer 0 as $u \to \infty$. Let $Z(u) = \mathbb{E}_Q[e^{RC_\tau} \mid C_0 = u]$. Then $Z(u)$ satisfies the renewal equation

$$Z(u) = \int_0^u Z(u - y) \, dB(y) + \int_u^\infty e^{-R(y-u)} \, dB(y).$$

It follows from Lemma C.14 that

$$\int_u^\infty e^{-R(y-u)}\,dB(y) = e^{Ru}\int_u^\infty e^{-Ry}\,dB(y)$$

is directly Riemann integrable and its integral is

$$\int_0^\infty\int_u^\infty e^{-R(y-u)}\,dB(y)\,dz = \frac{1}{R}\Bigl(1-\int_0^\infty e^{-Ry}\,dB(y)\Bigr).$$

Thus by the renewal theorem the limit of $Z(u)$ exists and is equal to

$$\frac{1}{R\int_0^\infty u\,dB(u)}\Bigl(1-\int_0^\infty e^{-Ry}\,dB(y)\Bigr).$$

Consider first $\int_0^\infty u\,dB(u)$. We find

$$\begin{aligned}\int_0^\infty u\,dB(u) &= -\mathbb{E}_Q[C_\tau \mid C_0=0] = -\mathbb{E}_{\mathbb{P}}[C_\tau e^{-RC_\tau}; \tau<\infty \mid C_0=0]\\ &= \frac{\lambda}{c}\int_0^\infty x e^{Rx}(1-G(x))\,dx = \frac{\lambda}{R^2c}(RM_Y'(R)-(M_Y(R)-1)).\end{aligned}$$

The integral in the numerator of the limit is

$$\int_0^\infty e^{-Ry}\,dB(y) = \mathbb{E}_Q[e^{RC_\tau} \mid C_0=0] = \mathbb{P}[\tau<\infty \mid C_0=0] = \frac{\lambda\mu}{c}.$$

Thus

$$\begin{aligned}\lim_{u\to\infty} Z(u) &= \frac{\frac{1}{R}\bigl(1-\frac{\lambda\mu}{c}\bigr)}{\frac{\lambda}{R^2c}(RM_Y'(R)-(M_Y(R)-1))}\\ &= \frac{c-\lambda\mu}{\lambda(M_Y'(R)-(M_Y(R)-1)/R)} = \frac{c-\lambda\mu}{\lambda M_Y'(R)-c},\end{aligned} \tag{8.1}$$

where we used the definition of the Lundberg exponent R. This proves the Cramér–Lundberg approximation (Theorem 5.7).

8.3 The Renewal Case

In Chap. 6 we only considered the process at the claim times. Then we can make the change of measure in discrete time, unless we look at Lundberg inequalities in finite time. However, in order to see how to proceed for the Ammeter risk model, we need to look for a continuous time martingale. As mentioned in Chap. 6 the process $\{C_t\}$ is not a Markov process. For our approach it is important to have the Markov property. Thus we need to Markovise the process. The future of the process depends on the time of the last claim. Let $W_t = t - T_{N_t}$ denote the time elapsed since the last claim. Then $\{(C_t, W_t)\}$ is a Markov process. Alternatively, we can consider $V_t = T_{N_t+1} - t$, the time left till the next claim. This seems a little strange because $\{V_t\}$ is not observable, but recall that in the Ammeter model we worked with a non-observable stochastic process, which was useful in order to obtain Lundberg's inequality in the general case. The stochastic process $\{(C_t, V_t)\}$ is also a Markov process, but note that the natural filtrations of $\{C_t\}$ and of $\{(C_t, V_t)\}$ are different.

8.3.1 Markovisation via the Time Since the Last Claim

In order to construct a martingale we have to assume that the interarrival time distribution F is absolutely continuous with density $F'(t)$. The next martingale follows from the theory of piecewise deterministic Markov processes, see [110].

Lemma 8.4 *Assume that $M_Y(r) < \infty$ such that the unique solution $\theta(r)$ to*

$$M_Y(r)M_T(-\theta - cr) = 1$$

exists (such a solution exists if $r \ge 0$). Then the process

$$L_t^r = \mathrm{e}^{-r(C_t-u)}\, M_Y(r)\frac{\mathrm{e}^{(\theta(r)+cr)W_t}}{1-F(W_t)}\int_{W_t}^{\infty} \mathrm{e}^{-(\theta(r)+cr)s}\, F'(s)\, \mathrm{d}s\, \mathrm{e}^{-\theta(r)t}$$

is a martingale.

Remark 8.5 Considering the above martingale at the times $0, T_1, \dots$ only yields the martingale obtained in Lemma 6.1. □

Proof We have $W_{T_{N_t+1}} = 0$, and $M_Y(r)\int_0^\infty \mathrm{e}^{-(\theta(r)+cr)s}\, F'(s)\, \mathrm{d}s = 1$. We also have that $T_{N_t+1} - t$ has density $F'(W_t + v)/(1 - F(W_t))$. Then

$$\begin{aligned}
\mathbb{E}[L^r_{T_{N_t+1}\wedge s} \mid \mathcal{F}_t] &= \frac{\int_0^{s-t} \mathrm{e}^{-r(C_t+cv-u)} M_Y(r)\mathrm{e}^{-\theta(r)(t+v)} F'(W_t+v)\, \mathrm{d}v}{1-F(W_t)} \\
&\quad + \frac{1-F(W_t+s-t)}{1-F(W_t)}\mathrm{e}^{-r(C_t+c(s-t)-u)} M_Y(r) \\
&\quad \times \frac{\mathrm{e}^{(\theta(r)+cr)(W_t+s-t)}}{1-F(W_t+s-t)}\int_{W_t+s-t}^{\infty} \mathrm{e}^{-(\theta(r)+cr)v}\, F'(v)\, \mathrm{d}v\mathrm{e}^{-\theta(r)s} \\
&= L_t^r\,,
\end{aligned}$$

where we replaced $W_t + v$ in the first integral with v. The same calculation gives

$$\mathbb{E}[L^r_{T_{N_t+k+1}\wedge s} \mid \mathcal{F}_t] = \mathbb{E}[\mathbb{E}[L^r_{T_{N_t+k+1}\wedge s} \mid \mathcal{F}_{T_{N_t+k}}] \mid \mathcal{F}_t] = L^r_t.$$

The family is uniformly integrable since $0 \le W_s \le W_t + s - t$. So we can let $k \to \infty$, which proves the martingale property. □

Define the measures for $A \in \mathcal{F}_t$

$$Q_r[A] = \mathbb{E}_{\mathbb{P}}[L^r_t; A].$$

Here it also turns out that the measure Q_r can be extended to the whole σ-algebra $\mathcal{F}$.

Lemma 8.6

(i) $\{(C_t, W_t)\}$ *is a renewal risk model under the measure* Q_r *with claim size distribution* $\tilde{G}(x) = M_T(-\theta(r) - cr)\int_0^x e^{ry}\, dG(y)$ *and interarrival time distribution* $\tilde{F}(t) = M_Y(r)\int_0^t e^{-(\theta(r)+cr)s}\, dF(s)$. *In particular,*

$$\lim_{t\to\infty} \frac{1}{t}(C_t - u) = -\theta'(r)$$

Q_r-a.s.

(ii) If $r \neq 0$ then Q_r and $\mathbb{P}$ are singular on the whole σ-algebra $\mathcal{F}$.

Proof (ii) is proved as in Lemma 8.3.

(i) Let $n \in \mathbb{N}$ and $A_1, A_2, \ldots, A_n, B_1, B_2, \ldots, B_n$ be Borel sets. For the moment we have to choose $A_1, \ldots, A_n$ such that T_n remains finite. This assumption can be relaxed when we have proved that the measure Q_r can be extended to $\mathcal{F}$. Then because $W_{T_n} = 0$

$$\begin{aligned}
&Q_r[T_1 \in A_1, T_2 - T_1 \in A_2, \ldots, T_n - T_{n-1} \in A_n, Y_1 \in B_1, Y_2 \in B_2, \ldots, Y_n \in B_n] \\
&\quad = \mathbb{E}_{\mathbb{P}}[L^r_{T_n}; T_1 \in A_1, T_2 - T_1 \in A_2, \ldots, T_n - T_{n-1} \in A_n, \\
&\qquad Y_1 \in B_1, Y_2 \in B_2, \ldots, Y_n \in B_n] \\
&\quad = \mathbb{E}_{\mathbb{P}}\Bigl[\exp\Bigl\{r\sum_{i=1}^n Y_i\Bigr\}\exp\Bigl\{-(\theta(r)+cr)\sum_{i=1}^n (T_i - T_{i-1})\Bigr\}; T_1 \in A_1, \\
&\qquad T_2 - T_1 \in A_2, \ldots, T_n - T_{n-1} \in A_n, Y_1 \in B_1, Y_2 \in B_2, \ldots, Y_n \in B_n\Bigr] \\
&\quad = \prod_{i=1}^n \mathbb{E}_{\mathbb{P}}[e^{rY_i}; Y_i \in B_i]\mathbb{E}_{\mathbb{P}}[e^{-(\theta(r)+cr)(T_i - T_{i-1})}; (T_i - T_{i-1}) \in A_i].
\end{aligned}$$

Thus $\{(C_t, W_t)\}$ is the renewal risk model with the parameters given by the assertion. A straightforward calculation yields

$$\lim_{t\to\infty} \frac{1}{t}(C_t - u) = c - \tilde{\lambda}\tilde{\mu} = -\theta'(r)$$

Q_r-a.s. □

The ruin probability can be expressed as

$$\begin{aligned}\psi(u) &= \mathbb{E}_{Q_r}\Bigl[e^{rC_\tau} \frac{(1 - F(W_\tau))e^{-(\theta(r)+cr)W_\tau}}{M_Y(r)\int_{W_\tau}^\infty e^{-(\theta(r)+cr)s} F'(s)\,ds} e^{\theta(r)\tau}; \tau < \infty\Bigr]e^{-ru} \\ &= \mathbb{E}_{Q_r}[e^{rC_\tau} e^{\theta(r)\tau}; \tau < \infty]e^{-ru} ,\end{aligned}$$

where we used that $W_\tau = 0$. As before we have to choose $r = R$ which we assume to exist. Then, because $\theta'(R) > 0$, the net profit condition is violated and

$$\psi(u) = \mathbb{E}_Q[e^{RC_\tau}]e^{-Ru}.$$

Lundberg's inequality (Theorem 6.6) follows immediately and the Cramér–Lundberg approximation (Theorem 6.8) can be proved in the same way as in the classical case.

8.3.2 Markovisation via the Time till the Next Claim

Choosing the Markovisation $\{(C_t, V_t)\}$ simplifies the situation. The martingale looks nicer and we do not have to assume that F is absolutely continuous. Let t_0 be fixed. Then the process is deterministic in the interval $(t_0, t_0 + V_{t_0})$, in fact $(C_t, V_t) = (C_{t_0} + c(t - t_0), V_{t_0} - (t - t_0))$. Because no event can happen the martingale must be constant in the interval $(t_0, t_0 + V_{t_0})$.

Lemma 8.7 *Assume that $M_Y(r) < \infty$ such that the unique solution $\theta(r)$ to*

$$M_Y(r)M_T(-\theta - cr) = 1$$

exists (such a solution exists if $r \ge 0$). Then the process

$$L_t^r = M_Y(r)e^{-r(C_t-u)}e^{-(\theta(r)+cr)V_t}e^{-\theta(r)t}$$

is a martingale.

Proof We have

$$\begin{aligned}\frac{\mathbb{E}[L^r_{T_{N_t+1}\wedge s} \mid \mathcal{F}_t]}{M_Y(r)} &= \frac{L^r_t \mathbb{I}_{V_t>s-t}}{M_Y(r)} \\ &\quad + \mathrm{e}^{-r(C_t-u)-\theta(r)t-(\theta(r)+cr)V_t} M_Y(r) M_T(-\theta(r)-cr)\mathbb{I}_{V_t\le s-t} \\ &= \frac{L^r_t}{M_Y(r)}\;,\end{aligned}$$

where we used that $V_{T_{N_t+1}}$ has the same distribution as T, independent of the other variables, and that $V_v + v$ remains constant on $v \in (t, t+V_t)$. By induction, we see that

$$\mathbb{E}[L^r_{T_{N_t+k}\wedge s} \mid \mathcal{F}_t] = L^r_t\,.$$

We note that $\exp\{r\sum_{\ell=N_t+1}^{N_s} Y_\ell\}$ and V_s are integrable, and that for $t+V_t \le v < T_{N_s}$, $V_v < \max\{s-t, V_{T_{N_s}}\}$ is bounded by an integrable random variable. Thus, the martingale is uniformly integrable, and the assertion follows. □

Note that $\mathbb{E}[L^r_{T_i} \mid \mathcal{F}^C_{T_i}]$, where $\{\mathcal{F}^C_t\}$ denotes the natural filtration of the risk process $\{C_t\}$, is the martingale considered in Chap. 6. Define as before $Q_r[A] = \mathbb{E}_{\mathbb{P}}[L^r_t; A]$ for $A \in \mathcal{F}_t$. From the lemma below it again follows that the measure can be extended to a measure on the whole σ-algebra $\mathcal{F}$. We again have

Lemma 8.8

(i) $\{(C_t, V_t)\}$ *is a renewal risk model under the measure* Q_r *with claim size distribution* $\tilde{G}(x) = M_T(-\theta(r)-cr)\int_0^x \mathrm{e}^{ry}\,\mathrm{d}G(y)$ *and interarrival distribution* $\tilde{F}(t) = M_Y(r)\int_0^t \mathrm{e}^{-(\theta(r)+cr)s}\,\mathrm{d}F(s)$. *In particular,*

$$\lim_{t\to\infty}\frac{1}{t}(C_t - u) = -\theta'(r)$$

Q_r-*a.s.*

(ii) If $r \neq 0$ *then* Q_r *and* $\mathbb{P}$ *are singular on the whole* σ-*algebra* $\mathcal{F}$.

Proof Left as an exercise. □

For the ruin probability we obtain

$$\begin{aligned}\psi(u) &= (M_Y(r))^{-1}\mathbb{E}_{Q_r}[\mathrm{e}^{rC_\tau}\mathrm{e}^{(\theta(r)+cr)V_\tau}\mathrm{e}^{\theta(r)\tau}; \tau<\infty]\mathrm{e}^{-ru} \\ &= \mathbb{E}_{Q_r}[\mathrm{e}^{rC_\tau}\mathrm{e}^{\theta(r)\tau}; \tau<\infty]\mathrm{e}^{-ru}\,.\end{aligned}$$

Note that V_τ is independent of $\mathcal{F}_{\tau-}$ and has the same distribution as $T_{n+1} - T_n$. Thus we again have to choose $r = R$. Lundberg's inequality (Theorem 6.6) and the Cramér–Lundberg approximation (Theorem 6.8) follow as before.

Another advantage of the second approach is that we can easily treat the general case, where T_1 has distribution F^1. This causes problems in the first approach, because it is hard to find the distribution of W_0 such that T_1 has the correct distribution. Such a distribution does not even exist in all the cases.

Change the martingale to

$$L_t^r = \frac{e^{-r(C_t-u)}e^{-(\theta(r)+cr)V_t}e^{-\theta(r)t}}{\mathbb{E}_{\mathbb{P}}[e^{-(\theta(r)+cr)V_0}]}.$$

Note that this includes the ordinary case. Then

$$\psi(u) = M_Y(r)\mathbb{E}_{\mathbb{P}}[e^{-(\theta(r)+cr)V_0}]\,\mathbb{E}_{Q_r}[e^{rC_\tau}e^{\theta(r)\tau}; \tau < \infty]e^{-ru}.$$

Lundberg's inequality (Theorem 6.6) is again trivial. For the Cramér–Lundberg approximation we have to show that $\mathbb{E}_Q[e^{RC_\tau}]$ converges as $u \to \infty$. Letting $Z(u) = \mathbb{E}_Q[e^{RC_\tau} \mid C_0 = u]$ and $Z^o(u) = \mathbb{E}_Q^o[e^{RC_\tau} \mid C_0 = u]$ be the corresponding quantity in the ordinary case, we have

$$Z(u) = \int_0^\infty \Bigl(\int_0^{u+ct} Z^o(u+ct-y)\,d\tilde{G}(y) + \int_{u+ct}^\infty e^{-R(y-u-ct)}\,d\tilde{G}(y)\Bigr)\,d\tilde{F}^1(t).$$

Because $Z^o(u)$ is bounded by 1 the Cramér–Lundberg approximation (6.7) follows from the bounded convergence theorem.

8.4 The Ammeter Risk Model

For the Ammeter risk model we already found the martingale (Lemma 7.10)

$$L_t^r = e^{-r(C_t-u)}e^{(\lambda_t(M_Y(r)-1)-cr-\theta(r))V_t}e^{-\theta(r)t},$$

where V_t is the time remaining till the next change of the intensity. With this martingale we define the new measure $Q_r[A] = \mathbb{E}_{\mathbb{P}}[L_t^r; A]$ for $A \in \mathcal{F}_t$.

In the classical case we obtained the new intensity $\tilde{\lambda} = \lambda M_Y(r)$. Thus it might be that under the new measure the process $\{\lambda_t\}$ is no longer the intensity process. Let us therefore denote the intensity under the measure Q_r by $\tilde{\lambda}_t$ and the intensity levels by $\tilde{L}_i$. From the lemma below it follows again that Q_r can be extended to the whole σ-algebra $\mathcal{F}_t$.

Lemma 8.9 *(i)* $\{(C_t, \tilde{\lambda}_t, V_t)\}$ *is an Ammeter risk model under the measure* Q_r *with claim size distribution* $\tilde{G}(x) = \int_0^x e^{ry}\,dG(y)/M_Y(r)$, *claim arrival intensity* $\tilde{\lambda}_t = M_Y(r)\lambda_t$, $\tilde{L}_i = M_Y(r)L_i$ *and*

$$\tilde{H}(\ell) = e^{-(cr+\theta(r))\Delta}\int_0^\ell e^{l\Delta(M_Y(r)-1)}\,dH(l).$$

In particular, $\mathbb{E}_{Q_r}[C_\Delta - u] = -\theta'(r)\Delta$.

(ii) If $r \neq 0$ then Q_r and P are singular on the whole σ-algebra $\mathcal{F}$.

Proof (ii) is proved as in Lemma 8.3.

(i) We first want to show that the processes $\{(C_t - C_{(i-1)\Delta} : (i-1)\Delta \le t < i\Delta)\}$ are independent and follow the same law. Let A_i be a set measurable with respect to

$$\sigma(L_i, N_{i\Delta} - N_{(i-1)\Delta}, T_{N_{(i-1)\Delta}+1}, \dots, T_{N_{i\Delta}}, Y_{N_{(i-1)\Delta}+1}, \dots, Y_{N_{i\Delta}}).$$

Then for $n \in \mathbb{N}$

$$\begin{aligned} Q_r[A_1, \dots, A_n] &= \mathbb{E}_{\mathbb{P}}[\mathrm{e}^{-r(C_{n\Delta}-u)}\mathrm{e}^{(L_{n+1}(M_Y(r)-1)-cr-\theta(r))\Delta}\mathrm{e}^{-\theta(r)n\Delta}; A_1, \dots, A_n] \\ &= \mathbb{E}_{\mathbb{P}}[\mathrm{e}^{-r(C_{n\Delta}-u)}\mathrm{e}^{-\theta(r)n\Delta}; A_1, \dots, A_n] \\ &= \prod_{i=1}^n \mathrm{e}^{-\theta(r)\Delta}\mathbb{E}_{\mathbb{P}}[\mathrm{e}^{-r(C_{i\Delta}-C_{(i-1)\Delta})}; A_i] = \prod_{i=1}^n Q_r[A_i]. \end{aligned}$$

Next we show that the claim sizes are iid., independent of $\{N_t\}$ and have the required distribution. Let $n \in \mathbb{N}$ and A be a $\sigma(L_1, T_1, \dots, T_n)$-measurable set. Let B_i be Borel sets. Then

$$\begin{aligned} Q_r[N_\Delta = n, A, Y_1 \in B_1, \dots, Y_n \in B_n] &= \mathrm{e}^{-\theta(r)\Delta}\mathbb{E}_{\mathbb{P}}[\mathrm{e}^{-r(C_\Delta-u)}; N_\Delta = n, A, Y_1 \in B_1, \dots, Y_n \in B_n] \\ &= \mathrm{e}^{-(cr+\theta(r))\Delta}\mathbb{P}[N_\Delta = n, A]\prod_{i=1}^n \mathbb{E}_{\mathbb{P}}[\mathrm{e}^{rY_i}; Y_i \in B_i]. \end{aligned}$$

Next we verify the distribution of L_1 and N_Δ. Let A be Borel measurable. Then

$$\begin{aligned} Q_r[L_1 \in A, N_\Delta = n] &= \mathrm{e}^{-(cr+\theta(r))\Delta}M_Y(r)^n \int_A \frac{(\ell\Delta)^n}{n!}\mathrm{e}^{-\ell\Delta}\,\mathrm{d}H(\ell) \\ &= \int_A \frac{(M_Y(r)\ell\Delta)^n}{n!}\mathrm{e}^{-M_Y(r)\ell\Delta}\mathrm{e}^{-(cr+\theta(r))\Delta}\mathrm{e}^{\ell\Delta(M_Y(r)-1)}\,\mathrm{d}H(\ell). \end{aligned}$$

Thus L_1 and N_Δ have the desired joint distribution. Lastly we have to show that given $N_\Delta = n$ the times $(T_1, \dots, T_n)$ have the same distribution as the order statistics of n iid. uniformly in $(0, \Delta)$ distributed random variables (Proposition C.3). That is, the same conditional distribution as under $\mathbb{P}$. Let $n \in \mathbb{N}$, A be a $\sigma(T_1, \dots, T_n)$-measurable set and B be a Borel set such that $\mathbb{P}[L_1 \in B] \neq 0$. Then

$$\begin{aligned} Q_r[A \mid L_1 \in B, N_\Delta = n] &= \frac{\mathrm{e}^{-(cr+\theta(r))\Delta}\mathbb{P}[A, L_1 \in B, N_\Delta = n]M_Y(r)^n}{\mathrm{e}^{-(cr+\theta(r))\Delta}\mathbb{P}[L_1 \in B, N_\Delta = n]M_Y(r)^n} \\ &= \mathbb{P}[A \mid L_1 \in B, N_\Delta = n] = \mathbb{P}[A \mid N_\Delta = n]. \end{aligned}$$

The last assertion follows straightforwardly noting that

$$\mathbb{E}_{Q_r}[C_\Delta - u] = (c - \mathbb{E}_{Q_r}[M_Y(r)L_i]\mathbb{E}_{Q_r}[Y_i])\Delta.$$

□

We now consider the ordinary case $F^1(\ell, v) = H(\ell)\mathbb{1}_{v \ge \Delta}$. As before it turns out to be convenient to choose $r = R$ and that then the net profit condition is violated. For the ruin probability we obtain

$$\psi(u) = \mathbb{E}_Q[\mathrm{e}^{RC_\tau}\mathrm{e}^{-(\lambda_\tau(M_Y(R)-1)-cR)V_\tau}]\,\mathrm{e}^{-Ru}.$$

The Lundberg inequality follows immediately noting that $\lambda_\tau > 0$ and therefore $\mathrm{e}^{-(\lambda_\tau(M_Y(R)-1)-cR)V_\tau} < \mathrm{e}^{cR\Delta}$. For the Cramér–Lundberg approximation we have a problem. How can we find the asymptotic distribution of $(C_\tau, \lambda_\tau, V_\tau)$ as $u \to \infty$? In order to use the approach that was successful in the Cramér–Lundberg and in the renewal case, we need to know this distribution. With a slightly modified renewal approach it is possible to prove the Cramér–Lundberg approximation, see [114] for details.

In order to find a lower bound for the ruin probability define $\tau_1 := \inf\{k\Delta : k \in \mathbb{N}, C_{k\Delta} < 0\}$, the first time when the process becomes negative at a regeneration time. We assume now that there exists an $r > R$ such that $M_L(\Delta(M_Y(r)-1)) < \infty$. Obviously $\mathbb{P}[\tau < \infty] \ge \mathbb{P}[\tau_1 < \infty]$ and

$$\begin{aligned}\mathbb{P}[\tau_1 < \infty] &= \mathbb{E}_Q[\mathrm{e}^{RC_{\tau_1}}\mathrm{e}^{-(\lambda_{\tau_1}(M_Y(R)-1)-cR)V_{\tau_1}}]\mathrm{e}^{-Ru}\\ &= \mathbb{E}_Q[\mathrm{e}^{RC_{\tau_1}}]\mathrm{e}^{-Ru} \ge \mathrm{e}^{R\mathbb{E}_Q[C_{\tau_1}]}\mathrm{e}^{-Ru}.\end{aligned}$$

Because $\mathbb{E}_Q[\mathrm{e}^{-(r-R)(C_t-u)}]$ exists, all moments of the increments of the random walk $C_{k\Delta} - u$ exist. Using the renewal theorem one can show that $\mathbb{E}_Q[C_{\tau_1}]$ converges to a finite value as $u \to \infty$.

The case with a general $F^1(\ell, v)$ is left to the reader as an exercise.

Bibliographical Remarks

The change of measure method is frequently used in the literature without mentioning the connection to martingales, see for instance [8, 9] or [10]. The technique used in this section can also be found in [11] or [110]. Importance sampling is discussed in [69]. Lemma 8.4 was first proved in [41, Thm. 10]. Lemma 8.7 was used in an early version of [40], but does not appear in the final version. For a proof, see also [55, p. 27].

Chapter 9
The Markov Modulated Risk Model

In the Ammeter model the claim arrival intensity was constant during the time interval $((k-1)\Delta, k\Delta)$. The value of Δ was taken to be one year. It would be preferable if the intensity was allowed to change more frequently. This can be achieved by choosing Δ to be smaller, but it would also be preferable if different intensity levels could hold for time intervals of different lengths. This is the model considered by Björk and Grandell [22]. However, to keep track of the time since the last change of the intensity or the time till the next intensity complicates considerations. If an intensity holds for an exponentially distributed time, the lack of memory property of the exponential distribution simplifies the approach. Such a model is the Markov modulated risk model. One first constructs a continuous time Markov chain on a finite state space. This Markov chain represents an environment process for the risk business. The environment determines the intensity level. Moreover, the claim size distribution may differ for different states of the environment. We will be able to generalise the results obtained for the classical risk model.

9.1 Definition of the Markov Modulated Risk Model

Let $\{J_t\}$ be a continuous time irreducible Markov chain with state space $(1, 2, \ldots, \mathcal{J})$ and *intensity matrix* $\boldsymbol{\eta} = (\eta_{ij})$, where $\mathcal{J} \in \mathbb{N} \setminus \{0\}$ and $\eta_{ij} \ge 0$ for $i \ne j$ and $\eta_{ii} = -\sum_{j \ne i} \eta_{ij} < 0$. This means that, if $J_t = i$, then $J_{t+s} = i$ for a $\text{Exp}(-\eta_{ii})$ distributed time and the next state will be $j \ne i$ with probability $\eta_{ij}/(-\eta_{ii})$. Then there is a unique stationary distribution $\boldsymbol{\pi}$ with the property $\boldsymbol{\pi\eta} = \mathbf{0}$. We have $\pi_i > 0$ for all $i \le \mathcal{J}$.

Let $\{L_i : i \le \mathcal{J}\}$ denote the intensity levels and let $\{G_i(y) : i \le \mathcal{J}\}$ be distribution functions of positive random variables. L_i will be the claim arrival intensity and $G_i(y)$

H. Schmidli, *Risk Theory*, Springer Actuarial,
DOI:10.1007/978-3-319-72005-0_9

will be the claim size distribution during the time where the environment process is in state i. For a formal definition let $\Sigma_0 = 0$ and for $n \in \mathbb{N}$

$$\Sigma_{n+1} = \inf\{t > \Sigma_n : J_t \neq J_{t-}\}$$

be the times when the environment changes. Let $\{C_t^{(i)}\}$ be independent Cramér–Lundberg models with initial capital 0 and premium rate c where the i-th model has claim arrival intensity L_i and claim size distribution $G_i(y)$. The claim number process of the i-th model is denoted by $\{N_t^{(i)}\}$. Let $C_{0-}^{(i)} = 0$, $N_{0-}^{(i)} = 0$, $N_{0-} = 0$ and $C_{0-} = u$. Define recursively

$$N_t := N_{\Sigma_n-} + N_t^{(J_t)} - N_{\Sigma_n-}^{(J_t)}, \qquad \Sigma_n \leq t < \Sigma_{n+1}$$

and

$$C_t := C_{\Sigma_n-} + C_t^{(J_t)} - C_{\Sigma_n-}^{(J_t)}, \qquad \Sigma_n \leq t < \Sigma_{n+1}\,.$$

$\{N_t\}$ is the number of claims of $\{C_t\}$ in the interval $(0, t]$. Moreover, $\{N_t\}$ is a Cox point process with intensity process $\lambda_t = L_{J_t}$. That is, given $\{J_t : t \geq 0\}$, $\{N_t\}$ is an inhomogeneous Poisson process with rate $\{\lambda_t\}$ and $\mathbb{E}[N_t \mid J_s, 0 \leq s \leq t] = \int_0^t \lambda_s \,\mathrm{d}s$. We denote by $0 = T_0 < T_1 < T_2 < \cdots$ the claim times, $T_k = \inf\{t : N_t \geq k\}$. A claim occurring at time t has distribution function $G_{J_t}(y)$.

For later use we will need some further notation. Let $\mu_i = \int_0^\infty (1 - G_i(y)) \,\mathrm{d}y$ be the expected value and $M_Y^{(i)}(r) = \int_0^\infty \mathrm{e}^{ry} \,\mathrm{d}G_i(y)$ be the moment generating function of a claim in state i. By $M_Y(r) = \max\{M_Y^{(i)}(r) : i \leq \mathcal{J}\}$ we denote the maximal value of the moment generating functions. Denote by $\boldsymbol{I}$ the identity matrix. For r such that $M_Y(r) < \infty$, let $\boldsymbol{S}(r)$ be the diagonal matrix with $S_{ii}(r) = L_i(M_Y^{(i)}(r) - 1)$ and $\boldsymbol{\Theta}(r) := \boldsymbol{\eta} + \boldsymbol{S}(r) - cr\boldsymbol{I}$. We will often have to condition on $\{J_0 = i\}$. We therefore denote by $\mathbb{P}_i[\cdot] = \mathbb{P}[\cdot \mid J_0 = i]$ the conditional measure and by $\mathbb{E}_i$ the corresponding expectation.

Definition 9.1 Let $\boldsymbol{A}$ be a square matrix. Then we denote by $\mathrm{e}^{t\boldsymbol{A}}$ the matrix exponential

$$\mathrm{e}^{t\boldsymbol{A}} := \sum_{n=0}^{\infty} \frac{t^n}{n!} \boldsymbol{A}^n\,.$$

Lemma 9.2 *For the Markov modulated risk model $\{C_t\}$ one has*

$$\mathbb{E}_i[\mathrm{e}^{-r(C_t-u)} \mathbb{I}_{\{J_t=j\}}] = (\mathrm{e}^{t\boldsymbol{Theta}(r)})_{ij}$$

provided $M_Y(r) < \infty$. In particular, $\mathbb{P}_i[J_t = j] = (\mathrm{e}^{t\boldsymbol{\eta}})_{ij}$.

Proof Let $f_{ij}(t) = \mathbb{E}_i[\mathrm{e}^{-r(C_t-u)} \mathbb{I}_{\{J_t=j\}}]$. We denote by $T = T_1 \wedge \Sigma_1$ the time of the first event. From $\mathbb{P}_i[T > t] = \mathbb{P}_i[\Sigma_1 > t, T_1 > t] = \mathrm{e}^{\eta_{ii}t}\mathrm{e}^{-L_i t}$, we see that T is exponentially distributed with parameter $L_i - \eta_{ii}$. Further, $\mathbb{P}_i[T \leq t, T_1 < \Sigma_1] = \int_0^t \mathrm{e}^{\eta_{ii}s} L_i \mathrm{e}^{-L_i s} \,\mathrm{d}s = L_i(1 - \mathrm{e}^{-(L_i-\eta_{ii})t})/(L_i - \eta_{ii})$. This means that, independent of T,

$T = T_1$ with probability $L_i/(L_i - \eta_{ii})$. The probability of no events in a small time interval $(0, h]$ is

$$\mathbb{P}_i[T > h] = \mathrm{e}^{-(L_i-\eta_{ii})h} = 1 - (L_i - \eta_{ii})h + o(h) .$$

The probability of exactly one event is

$$\int_0^h L_i \mathrm{e}^{-(L_i-\eta_{ii})t}\mathrm{e}^{-(L_i-\eta_{ii})(h-t)}\,\mathrm{d}t + \sum_{\substack{j=1\\ j\neq i}}^{\mathcal{J}} \int_0^h \eta_{ij}\mathrm{e}^{-(L_i-\eta_{ii})t}\mathrm{e}^{-(L_j-\eta_{jj})(h-t)}\,\mathrm{d}t$$
$$= (L_i - \eta_{ii})h + o(h) .$$

This implies that the probability of two or more events in $(0, h]$ is $o(h)$. Thus, as $h \downarrow 0$

$$\mathbb{P}[N_{t+h} - N_t = 0 \mid J_t = k] = 1 - L_k h + o(h) ,$$
$$\mathbb{P}[N_{t+h} - N_t = 1 \mid J_t = k] = L_k h + o(h)$$

and

$$\mathbb{P}[J_{t+h} = \ell \mid J_t = k] = I_{k\ell} + \eta_{k\ell} h + o(h) .$$

Therefore, for h small,

$$\begin{aligned} f_{ij}(t+h) = f_{ij}(t)(1 + \eta_{jj}h + o(h))\big\{&(1 - L_j h + o(h))\mathrm{e}^{-rch} \\ &+ (L_j h + o(h))\mathrm{e}^{-rch} M_Y^{(j)}(r)\big\} \\ &+ \sum_{\substack{k=1\\ k\neq j}}^{\mathcal{J}} f_{ik}(t)(\eta_{kj}h + o(h))\mathrm{e}^{-rch} + o(h) \end{aligned}$$
$$= f_{ij}(t)\mathrm{e}^{-rch} + f_{ij}(t)L_j h(M_Y^{(j)}(r) - 1) + \sum_{k=1}^{\mathcal{J}} f_{ik}(t)\eta_{kj}h + o(h) .$$

In particular, $f_{ij}(t)$ is right continuous. Replacing t by $t - h$ shows left continuity. Reordering the terms yields

$$\frac{f_{ij}(t+h) - f_{ij}(t)}{h}$$
$$= -\frac{1 - \mathrm{e}^{-rch}}{h} f_{ij}(t) + f_{ij}(t)L_j(M_Y^{(j)}(r) - 1) + \sum_{k=1}^{\mathcal{J}} f_{ik}(t)\eta_{kj} + o(1) ,$$

and the analogous equation with t replaced by $t-h$. Thus, letting $h \downarrow 0$, $f_{ij}(t)$ satisfies the differential equation

$$f'_{ij}(t) = -crf_{ij}(t) + f_{ij}(t)L_j(M_Y^{(j)}(r) - 1) + \sum_{k=1}^{\mathcal{J}} f_{ik}(t)\eta_{kj} = \sum_{k=1}^{\mathcal{J}} f_{ik}(t)\Theta_{kj}(r)$$

with the initial condition $f_{ij}(0) = I_{ij}$. The only solution to the above system of differential equations is $f_{ij}(t) = (\mathrm{e}^{t\boldsymbol{\Theta}(r)})_{ij}$. The second assertion follows for $r = 0$. □

Note that the stationary distribution satisfies

$$\boldsymbol{\pi}\mathrm{e}^{t\eta} = \boldsymbol{\pi} .$$

Let for the moment $u = 0$ and let $V_i(t) = \int_0^t \mathbb{I}_{\{J_s=i\}} \,\mathrm{d}s$ denote the time $\{J_s\}$ has spent in state i. Then

$$\frac{C_t}{t} = \sum_{i=1}^{\mathcal{J}} \frac{V_i(t)}{t} \frac{\int_0^t \mathbb{I}_{\{J_s=i\}} \,\mathrm{d}C_s}{V_i(t)} ,$$

where

$$\int_0^t \mathbb{I}_{\{J_s=i\}} \,\mathrm{d}C_s = \int_0^t \mathbb{I}_{\{J_s=i\}} c \,\mathrm{d}s - \sum_{j=1}^{N_t} \mathbb{I}_{\{J_{T_j}=i\}} Y_j .$$

It is well-known that $V_i(t)/t \to \pi_i$. The random variable $\int_0^t \mathbb{I}_{\{J_s=i\}} \,\mathrm{d}C_s$ has the same distribution as $C_{V_i(t)}^{(i)}$ and it follows from the classical model that $C_{V_i(t)}^{(i)}/V_i(t)$ tends to $c - L_i\mu_i$ a.s. Thus C_t/t converges a.s. to $c - \sum_{i=1}^{\mathcal{J}} \pi_i L_i \mu_i$. In order to avoid $\psi(u) = 1$ for all u the latter limit should be strictly positive. This means the net profit condition reads

$$c > \sum_{i=1}^{\mathcal{J}} \pi_i L_i \mu_i .$$

9.2 The Lundberg Exponent and Lundberg's Inequality

The matrix $\mathrm{e}^{\boldsymbol{\Theta}(r)}$ has only strictly positive elements by Lemma 9.2. Let $\mathrm{e}^{\theta(r)} = \operatorname{spr} \mathrm{e}^{\boldsymbol{\Theta}(r)}$ denote the largest absolute value of the eigenvalues. By the Frobenius theorem $\mathrm{e}^{\theta(r)}$ is an eigenvalue of $\mathrm{e}^{\boldsymbol{\Theta}(r)}$, it is the only algebraic eigenvalue with absolute value $\operatorname{spr} \mathrm{e}^{\boldsymbol{\Theta}(r)}$ and its eigenvector $\boldsymbol{g}(r)$ consists of strictly positive elements. Moreover, $\boldsymbol{g}(r)$ is the only non-trivial eigenvector consisting of positive elements only. It follows in particular that $\boldsymbol{g}(0) = \mathbf{1} := (1, 1, \dots, 1)^\top$ because $\boldsymbol{\eta}\mathbf{1} = \mathbf{0}$. We normalise $\boldsymbol{g}(r)$ so that $\boldsymbol{\pi}\boldsymbol{g}(r) = 1$. We find the following martingale.

Lemma 9.3 *The stochastic process*

$$L_t^r = g_{J_t}(r)/\mathbb{E}[g_{J_0}(r)]\,\mathrm{e}^{-r(C_t-u)-\theta(r)t}$$

is a martingale. Moreover, the function $\theta(r)$ is convex and

$$\theta'(0) = -\Bigl(c - \sum_{i=1}^{\mathcal{J}} \pi_i L_i \mu_i\Bigr) < 0\,.$$

Proof In order to simplify the notation we write g_i instead of $g_i(r)$. Note that

$$\mathbb{E}_i[g_{J_t} \mathrm{e}^{-r(C_t-u)}] = (\mathrm{e}^{t\boldsymbol{\Theta}(r)} g)_i = \mathrm{e}^{\theta(r)t} g_i\,.$$

Because the increments of $\{C_{t+s} : s \ge 0\}$ depend on $\mathcal{F}_t$ via J_t only, the martingale property is proved.

It follows from Lemma 1.13 that $\log \sum_{i=1}^{\mathcal{J}} \mathbb{E}_i[\mathrm{e}^{-r(C_t-u)}; J_t = j]$ is strictly convex. It is clear that

$$\limsup_{n\to\infty}((\mathrm{e}^{\boldsymbol{\Theta}(r)})^n \mathbf{1})_i^{1/n} \le \mathrm{e}^{\theta(r)}\,.$$

Let $\bar{g} = \max\{g_i\}$. Then

$$\begin{aligned}\liminf_{n\to\infty}((\mathrm{e}^{\boldsymbol{\Theta}(r)})^n \mathbf{1})_i^{1/n} &= \liminf_{n\to\infty}((\mathrm{e}^{\boldsymbol{\Theta}(r)})^n \bar{g}\mathbf{1})_i^{1/n} \ge \liminf_{n\to\infty}((\mathrm{e}^{\boldsymbol{\Theta}(r)})^n g)_i^{1/n} \\ &= \mathrm{e}^{\theta(r)}\,.\end{aligned}$$

Thus there exists a series $\{f_n(r) : n \in \mathbb{N}\}$ of strictly convex functions converging to $\theta(r)$. Let $r_1 < r_2$ such that $M_Y(r_2) < \infty$. Let $\alpha \in (0, 1)$. Then

$$\begin{aligned}\theta(\alpha r_1 + (1-\alpha)r_2) &= \lim_{n\to\infty} f_n(\alpha r_1 + (1-\alpha)r_2) \\ &\le \lim_{n\to\infty} \alpha f_n(r_1) + (1-\alpha) f_n(r_2) = \alpha\theta(r_1) + (1-\alpha)\theta(r_2)\end{aligned}$$

and $\theta(r)$ is convex.

Note that $\theta(r)$ is an eigenvalue of $\boldsymbol{\Theta}(r)$, i.e. a solution to the equation $\det(\boldsymbol{\Theta}(r) - \theta \boldsymbol{I}) = 0$. By the implicit function theorem $\theta(r)$ is differentiable and therefore $\boldsymbol{g}(r)$ is differentiable, too. We have $\boldsymbol{\pi\Theta}(r)\boldsymbol{g}(r) = \theta(r)\boldsymbol{\pi g}(r) = \theta(r)$. In particular,

$$\theta'(r) = \boldsymbol{\pi\Theta}'(r)\boldsymbol{g}(r) + \boldsymbol{\pi\Theta}(r)\boldsymbol{g}'(r)\,.$$

Letting $r = 0$ the result for $\theta'(0)$ follows noting that $\boldsymbol{\pi\Theta}(0) = \boldsymbol{\pi\eta} = \mathbf{0}$ and $\boldsymbol{g}(0) = \mathbf{1}$. □

Because $\theta(r)$ is convex there might exist a second solution $R \ne 0$ to the equation $\theta(r) = 0$. If such a solution exists then $R > 0$ and the solution is unique. We again call it the *Lundberg exponent* or *adjustment coefficient*. Let us now assume that R exists. As in Chap. 8 let us define the new measure $Q_r[A] = \mathbb{E}_{\mathbb{P}}[L_t^r; A]$. It will again follow from the lemma below that Q_r can be extended to $\mathcal{F}$.

Lemma 9.4 *Under the measure Q_r, $\{(C_t, J_t)\}$ is a Markov modulated risk model with intensity matrix $\tilde{\eta}_{ij} = (g_i(r))^{-1} g_j(r)\eta_{ij}$ if $i \neq j$, claim size distributions $\tilde{G}_i(x) = \int_0^x e^{ry} \,dG_i(y)/M_Y^{(i)}(r)$ and claim arrival intensities $\tilde{L}_i = L_i M_Y^{(i)}(r)$. In particular,*

$$\lim_{t\to\infty} \frac{1}{t}(C_t - u) = -\theta'(r)$$

Q_r-a.s.

Proof Let B_k be $\sigma(\Sigma_k - \Sigma_{k-1}, N_{\Sigma_k} - N_{\Sigma_{k-1}}, T_j - T_{j-1}, Y_j; N_{\Sigma_{k-1}} < j \leq N_{\Sigma_k})$-measurable sets, $n \in \mathbb{N}$ and $i_k \leq \mathcal{J}$. Before we have proved that Q_r can be extended to $\mathcal{F}$ we have to choose sets B_k such that we only consider the process in finite time. Let $i_k \in (1, \ldots, \mathcal{J})$. Then

$$\begin{aligned}
&Q_r[J_{\Sigma_{k-1}} = i_k, J_{\Sigma_n} = i_{n+1}, B_k, 1 \leq k \leq n] \\
&= \frac{g_{i_{n+1}}}{\mathbb{E}_{\mathbb{P}}[g_{J_0}]} \mathbb{E}_{\mathbb{P}}\Bigg[\prod_{k=1}^{n} e^{-r(C_{\Sigma_k} - C_{\Sigma_{k-1}})} e^{-\theta(r)(\Sigma_k - \Sigma_{k-1})}; \\
&\qquad\qquad J_{\Sigma_{k-1}} = i_k, J_{\Sigma_n} = i_{n+1}, B_k, 1 \leq k \leq n\Bigg] \\
&= \frac{g_{i_{n+1}}}{\mathbb{E}_{\mathbb{P}}[g_{J_0}]} \mathbb{E}_{\mathbb{P}}[e^{-r(C_{\Sigma_1} - u)} e^{-\theta(r)\Sigma_1}; J_0 = i_1, B_1] \\
&\quad \times \prod_{k=2}^{n} \mathbb{E}_{\mathbb{P}}[e^{-r(C_{\Sigma_k} - C_{\Sigma_{k-1}})} e^{-\theta(r)(\Sigma_k - \Sigma_{k-1})}; J_{\Sigma_{k-1}} = i_k, B_k \mid J_{\Sigma_{k-2}} = i_{k-1}] \\
&\quad \times \mathbb{P}[J_{\Sigma_n} = i_{n+1} \mid J_{\Sigma_{n-1}} = i_n]
\end{aligned}$$

and therefore the process in the interval $[\Sigma_k, \Sigma_{k+1})$ depends on $\mathcal{F}_{\Sigma_k-}$ via $J_{\Sigma_{k-1}}$ only. In particular,

$$\begin{aligned}
&Q_r[J_{\Sigma_k} = i_{k+1}, B_k, 1 \leq k \leq n \mid J_0 = i_1] \\
&= \frac{g_{i_{n+1}}}{g_{i_1}} \mathbb{E}_{\mathbb{P}}[e^{-r(C_{\Sigma_1} - u)} e^{-\theta(r)\Sigma_1}; B_1 \mid J_0 = i_1] \\
&\quad \times \prod_{k=2}^{n} \mathbb{E}_{\mathbb{P}}[e^{-r(C_{\Sigma_k} - C_{\Sigma_{k-1}})} e^{-\theta(r)(\Sigma_k - \Sigma_{k-1})}; J_{\Sigma_{k-1}} = i_k, B_k \mid J_{\Sigma_{k-2}} = i_{k-1}] \\
&\quad \times \mathbb{P}[J_{\Sigma_n} = i_{n+1} \mid J_{\Sigma_{n-1}} = i_n]
\end{aligned}$$

and

$$\begin{aligned}
&Q_r[B_2 \mid J_0 = i_1, J_{\Sigma_1} = i_2] \\
&\qquad = \mathbb{E}_{\mathbb{P}}[e^{-r(C_{\Sigma_2} - C_{\Sigma_1})} e^{-\theta(r)(\Sigma_2 - \Sigma_1)}; B_2 \mid J_0 = i_1, J_{\Sigma_1} = i_2] \\
&\qquad = \mathbb{E}_{\mathbb{P}}[e^{-r(C_{\Sigma_2} - C_{\Sigma_1})} e^{-\theta(r)(\Sigma_2 - \Sigma_1)}; B_2 \mid J_{\Sigma_1} = i_2] = Q_r[B_2 \mid J_{\Sigma_1} = i_2]
\end{aligned}$$

and therefore the process in the interval $[\Sigma_n, \Sigma_{n+1})$ depends on $\mathcal{F}_{\Sigma_n}$ via J_{Σ_n} only. Next we compute the distribution of Σ_1. We have

$$\begin{aligned} Q_r[\Sigma_1 > v \mid J_0 = i] &= \mathbb{E}_{\mathbb{P}}[\mathrm{e}^{-r(C_v-u)-\theta(r)v}; \Sigma_1 > v \mid J_0 = i] \\ &= \mathrm{e}^{(L_i(M_Y^{(i)}(r)-1)-cr-\theta(r))v}\mathrm{e}^{\eta_{ii}v} \end{aligned}$$

and therefore Σ_1 is $\text{Exp}(-(\boldsymbol{\Theta}(r) - \theta(r)\boldsymbol{I})_{ii})$ distributed. It follows that $\{J_t\}$ is a continuous time Markov chain under Q_r. In order to find the intensity we compute

$$\begin{aligned} &Q_r[J_{\Sigma_1} = j \mid J_0 = i] \\ &\quad = \frac{g_j \mathbb{P}[J_{\Sigma_1} = j \mid J_0 = i]}{g_i}\mathbb{E}_{\mathbb{P}}[\mathrm{e}^{(L_i(M_Y^{(i)}(r)-1)-cr-\theta(r))\Sigma_1} \mid J_0 = i] \\ &\quad = \frac{g_j \eta_{ij}}{-g_i(\boldsymbol{\Theta}(r) - \theta(r)\boldsymbol{I})_{ii}} \end{aligned}$$

and the desired intensity matrix under Q_r follows. Note that we could have considered $Q_r[J_{\Sigma_1} = j \mid J_0 = i, \Sigma_1 \le t]$, which would have given the same result.

Let $n \in \mathbb{N}$, A be a $\sigma(T_1, \dots, T_n)$-measurable set and B_k be Borel sets. Then

$$\begin{aligned} &Q_r[\Sigma_1 > v, N_v = n, A, Y_k \in B_k, 1 \le k \le n \mid J_0 = i] \\ &\quad = \mathrm{e}^{-(cr+\theta(r))v}\mathbb{P}[\Sigma_1 > v, N_v = n, A \mid J_0 = i]\prod_{k=1}^n \int_{B_k} \mathrm{e}^{ry}\,\mathrm{d}G_i(y) \end{aligned}$$

and thus the claim size distribution is as desired. Moreover,

$$Q_r[A \mid N_v = n, \Sigma_1 > v, J_0 = i] = \mathbb{P}[A \mid N_v = n, \Sigma_1 > v, J_0 = i]\;,$$

from which it follows that, conditioned on the number of claims, the claim times are distributed as in a Poisson process. Finally,

$$Q_r[N_v = n \mid \Sigma_1 > v, J_0 = i] = \frac{(L_i M_Y^{(i)}(r)v)^n}{n!}\mathrm{e}^{-L_i M_Y^{(i)}(r)v}\;,$$

which shows that $\{(C_t, J_t)\}$ is a Markov modulated risk model.

Under the measure Q_r the process

$$\tilde{L}_t^s = \frac{g_{J_t}(r+s)}{g_{J_t}(r)}\mathrm{e}^{-s(C_t-u)}\mathrm{e}^{-(\theta(s+r)-\theta(r))t}$$

is a martingale because for $t \ge v$

$$\begin{aligned}
&\mathbb{E}_{Q_r}\Bigl[\frac{g_{J_t}(r+s)}{g_{J_t}(r)}\mathrm{e}^{-s(C_t-u)-(\theta(s+r)-\theta(r))t} \Bigm| \mathcal{F}_v\Bigr] \\
&\quad = \frac{\mathbb{E}_{\mathbb{P}}\Bigl[\frac{g_{J_t}(r+s)}{g_{J_t}(r)}\mathrm{e}^{-s(C_t-u)-(\theta(s+r)-\theta(r))t}\frac{g_{J_t}(r)}{\mathbb{E}_{\mathbb{P}}[g_{J_0}(r)]}\mathrm{e}^{-r(C_t-u)-\theta(r)t} \Bigm| \mathcal{F}_v\Bigr]}{\frac{g_{J_v}(r)}{\mathbb{E}_{\mathbb{P}}[g_{J_0}(r)]}\mathrm{e}^{-r(C_v-u)-\theta(r)v}} \\
&\quad = \mathbb{E}_{\mathbb{P}}\Bigl[\frac{g_{J_t}(r+s)}{g_{J_v}(r)}\mathrm{e}^{-(r+s)(C_t-u)-\theta(r+s)t} \Bigm| \mathcal{F}_v\Bigr]\mathrm{e}^{r(C_v-u)+\theta(r)v} \\
&\quad = \frac{g_{J_v}(r+s)}{g_{J_v}(r)}\mathrm{e}^{-(r+s)(C_v-u)-\theta(r+s)v}\mathrm{e}^{r(C_v-u)+\theta(r)v} = \tilde{L}_v^s\,.
\end{aligned}$$

As in Lemma 9.3 it follows that

$$\lim_{t\to\infty}\frac{1}{t}(C_t-u) = -\frac{\mathrm{d}}{\mathrm{d}s}(\theta(r+s)-\theta(r))|_{s=0} = -\theta'(r)\,.$$

□

Let again τ denote the time of ruin and $\psi(u)$ denote the ruin probability. If we write $\psi(u)$ in terms of the measure Q_r we obtain

$$\psi(u) = \mathbb{E}_{\mathbb{P}}[g_{J_0}(r)]\mathbb{E}_{Q_r}[(g_{J_\tau}(r))^{-1}\mathrm{e}^{rC_\tau}\mathrm{e}^{\theta(r)\tau};\tau<\infty]\mathrm{e}^{-ru}\,. \tag{9.1}$$

The latter expression is only simpler than $\mathbb{P}[\tau<\infty]$ if $r=R$. In the sequel we write Q instead of Q_R and g_i instead of $g_i(R)$.

We can now prove Lundberg's inequality.

Theorem 9.5 *Let $\{(C_t, J_t)\}$ be a Markov modulated risk model and assume that R exists. Let $\{g_i : 1 \le i \le \mathcal{J}\}$ be as above and $g_{\min} = \min\{g_i : 1 \le i \le \mathcal{J}\}$. Then*

$$\psi(u) < \mathbb{E}_{\mathbb{P}}[g_{J_0}]/g_{\min}\,\mathrm{e}^{-Ru}\,.$$

Proof Note that $g_{\min} > 0$. Thus the assertion follows readily from (9.1), noting that $\mathrm{e}^{RC_\tau} < 1$. □

9.3 The Cramér–Lundberg Approximation

When proving the Cramér–Lundberg approximation in the classical case (8.1) and in the renewal case there was no problem in using the renewal equation because each ladder time was a regeneration point. This is no longer the case for the Markov modulated risk model. We therefore have to find a slightly different version of the ladder epochs.

Assume that $L_1 \neq 0$ and that $J_0 = 1$. Let

$$\tau_- := \inf\{t > 0 : J_t = 1 \text{ and } C_t = \inf_{0 \le s \le t} C_s\}\,.$$

Because of the lack of memory property of the exponential distribution, τ_- is a regeneration point.

Lemma 9.6 *We have $Q[\tau_- < \infty] = 1$.*

Proof Let $S_0 = \inf\{t > 0 : C_t < 0\}$ and $S_{n+1} = \inf\{t > S_n : C_t < C_{S_n}\}$. Note that under Q, $C_t \to -\infty$, and therefore all S_k are finite. Let

$$A_n = \{J_{S_n+1} = 1, J_t = 1 \text{ for } t \in (S_n+1, S_n+2), N_{S_n+1} - N_{S_n} = 0, \\ N_{S_n+2} - N_{S_n+1} \ge 1, Y_{N_{S_n+1}+1} > 2c\}\,.$$

Then $\kappa_i = \mathbb{P}[A_n \mid J_{S_n} = i] > 0$ for all i. This implies that $\sum_{n=1}^\infty \mathbb{P}[A_n \mid \mathcal{F}_{S_n}] = \infty$. By the generalised Borel–Cantelli lemma B.4, infinitely many A_n occur. This implies that $\tau_- < \infty$. □

If $C_{\tau_-} \ge 0$ then ruin has not yet occurred. Denote by $B(x) = Q[u - C_{\tau_-} \le x]$ the modified ladder height distribution under Q. Thus

$$Z(u) = \mathbb{E}_Q[(g_{J_\tau})^{-1} e^{RC_\tau} \mid C_0 = u, J_0 = 1]$$

satisfies the renewal equation

$$Z(u) = \int_0^u Z(u-y)\,\mathrm{d}B(y) + z(u)\,,$$

where

$$z(u) = \mathbb{E}_Q[(g_{J_\tau})^{-1} e^{RC_\tau}; C_{\tau_-} < 0 \mid C_0 = u, J_0 = 1]\,.$$

In order to show that $Z(u)$ converges as $u \to \infty$ we have to show that $z(u)$ is directly Riemann integrable. To avoid $\lim_{u\to\infty} Z(u) = 0$ we assume that there exists an $r > R$ such that $M_Y(r) < \infty$.

Let us first show that $z(u)$ is a continuous function. Define

$$z_i(u) = \mathbb{E}_Q[(g_{J_\tau})^{-1} e^{RC_\tau}; C_{\tau_-} < 0 \mid C_0 = u, J_0 = i]\,.$$

Note that $z_i(u)$ is bounded. Then for $h > 0$ small we condition on the time $T \wedge h$ where $T = T_1 \wedge \Sigma_1$ and T_1 is the time of the first claim. Recall that T is exponentially distributed with parameter $\tilde{L}_1 - \tilde{\eta}_{11}$ under the measure Q. Then

$$z(u) = e^{-(\tilde{L}_1-\tilde{\eta}_{11})h} z(u+ch) + \int_0^h \int_0^\infty z(u+ct-y)\, d\tilde{G}_1(y)\tilde{L}_1 e^{-(\tilde{L}_1-\tilde{\eta}_{11})t}\, dt$$
$$+ \sum_{j=2}^{\mathcal{J}} \int_0^h z_j(u+ct)\tilde{\eta}_{1j} e^{-(\tilde{L}_1-\tilde{\eta}_{11})t}\, dt\,.$$

Letting $h \to 0$ shows that $z(u)$ is right continuous. Replacing u by $u - ch$ gives

$$z(u-ch) = e^{-(\tilde{L}_1-\tilde{\eta}_{11})h} z(u) + \sum_{j=2}^{\mathcal{J}} \int_0^h z_j(u-c(h-t))\tilde{\eta}_{1j} e^{-(\tilde{L}_1-\tilde{\eta}_{11})t}\, dt$$
$$+ \int_0^h \int_0^\infty z(u-c(h-t)-y)\, d\tilde{G}_1(y)\tilde{L}_1 e^{-(\tilde{L}_1-\tilde{\eta}_{11})t}\, dt$$

and $z(u)$ is left continuous too.

It remains to show that $z(u)$ has a directly Riemann integrable upper bound. Then it will follow from Lemma C.13 that $z(u)$ is directly Riemann integrable. It is clear that

$$g_{\min} z(u) \le Q[C_{\tau_-} < 0 \mid C_0 = u, J_0 = 1] = Q[-C_{\tau_-} > u \mid J_0 = 1, C_0 = 0]\,.$$

The latter function is monotone. Thus it is enough to show that it is integrable. We have

$$\int_0^\infty Q[-C_{\tau_-} > u \mid J_0 = 1, C_0 = 0]\, du = \mathbb{E}_Q[-C_{\tau_-} \mid J_0 = 1, C_0 = 0] < \infty\,.$$

The latter follows from the Wiener–Hopf factorisation because

$$\mathbb{E}_Q[e^{(r-R)/2\,(-C_{\tau_-})} \mid J_0 = 1, C_0 = 0]$$
$$= \mathbb{E}_{\mathbb{P}}[e^{(r+R)/2\,(-C_{\tau_-})}; \tau_- < \infty \mid J_0 = 1, C_0 = 0] < \infty\,,$$

since $\mathbb{E}_{\mathbb{P}}[e^{-rC_t}] < \infty$ for all $t \ge 0$. It follows readily that for $\tau^* = \inf\{T_n : n \ge 0, J_{T_n} = 1\}$ we also have $\mathbb{E}_{\mathbb{P}}[\exp\{-rC_{\tau^*}\}] < \infty$. We have just shown that $z(u)$ is directly Riemann integrable and therefore that the limit of $Z(u)$ exists as $u \to \infty$.

Theorem 9.7 *Let $\{(C_t, J_t)\}$ be a Markov modulated risk model and let $\{g_i : 1 \le i \le \mathcal{J}\}$ as above. Assume that R exists and that there is an $r > R$ such that $M_Y(r) < \infty$. Then*

$$\lim_{u\to\infty} \psi(u) e^{Ru} = \mathbb{E}_{\mathbb{P}}[g_{J_0}]\, C$$

for some constant $C > 0$ which is independent of the distribution of J_0.

Proof We have already seen that the assertion is true if $J_0 = 1$ a.s. Denote the limit of $Z(u)$ by C. Let $\tilde{\tau} = \inf\{t \ge 0 : J_t = 1\}$ and $B_1(x) = Q[u - C_{\tilde{\tau}} \le x]$. Denote by

$$\tilde{Z}(u) = \mathbb{E}_Q[(g_{J_\tau})^{-1} e^{RC_\tau} \mid C_0 = u]$$

the analogue of $Z(u)$. Then

$$\tilde{Z}(u) = \int_{-\infty}^{u} Z(u - y) \, dB_1(y) + \tilde{z}(u) ,$$

where

$$\tilde{z}(u) = \mathbb{E}_Q[(g_{J_\tau})^{-1} e^{RC_\tau}; \tau \le \tilde{\tau} \mid C_0 = u] .$$

Because $Q[\tau \le \tilde{\tau} \mid C_0 = u]$ tends to 0 as $u \to \infty$ we find that $\tilde{z}(u)$ tends to 0 as $u \to \infty$. Because $Z(x)$ is a bounded function we can interchange the limit and the integration to find that

$$\lim_{u\to\infty} \tilde{Z}(u) = \int_{-\infty}^{\infty} \lim_{u\to\infty} Z(u - y) \, dB_1(y) = C .$$

□

9.4 Subexponential Claim Sizes

Assume that one or several of the claim size distributions G_i are subexponential. Assume that $G_1(x)$ has the heaviest tail of all claim size distributions, more specifically, there exists constants $0 \le K_i < \infty$ such that

$$\lim_{y\to\infty} \frac{1 - G_i(y)}{1 - G_1(y)} = K_i$$

for all $i \le \mathcal{J}$. We assume without further mention that $L_1 > 0$. The following result is much more complicated to prove than the corresponding result in the renewal risk model. We therefore state the result without proof.

Theorem 9.8 *Assume that*

$$\lim_{y\to\infty} \frac{1 - G_i(y)}{1 - G_1(y)} = K_i < \infty ,$$

that $L_1 > 0$ *and that*

$$\frac{1}{\mu_1} \int_0^x (1 - G_1(y)) \, dy$$

is subexponential. Let $K = \sum_i \pi_i L_i K_i$. *Then*

$$\lim_{u\to\infty} \frac{\psi(u)}{\int_u^\infty (1 - G_1(y))\,\mathrm{d}y} = \frac{K}{c - \sum_i \pi_i L_i \mu_i}$$

independent of the distribution of the initial state. □

Let us consider the problem intuitively. Assume the process starts in state 1. Consider the random walk generated by the process by considering only the time points when the process J_t jumps to state 1. The increments of this random walk have a left tail that is asymptotically comparable with the tail of G_1 (only the heaviest tail contributes). Thus it follows, as in the renewal case, that the ruin probability of the random walk has the desired behaviour.

The time between two regeneration points has a light tailed distribution. The difference between the infimum of the random walk and the infimum of the risk process is smaller than c times the distance between the corresponding two regeneration points. The infimum of the random walk has a tail proportional to $\int_u^\infty (1 - G_1(y))\,\mathrm{d}y$, which will asymptotically dominate the light tail of the distance between two regeneration points. Thus the tail of the distribution of the infimum of the random walk and tail of the distribution of the infimum of the risk process are asymptotically equivalent.

9.5 Finite Time Lundberg Inequalities

Let $0 \le \underline{y} < \overline{y} < \infty$. As for the other models we want to consider probabilities of the form $\mathbb{P}[\underline{y}u < \tau \le \overline{y}u]$. We can proceed exactly as in the classical case (Sect. 5.14)

$$\begin{aligned}
\mathbb{E}_{\mathbb{P}}[g_{J_0}(r)]\mathrm{e}^{-ru} &= \mathbb{E}_{\mathbb{P}}[g_{J_{\tau\wedge\overline{y}u}}(r)\mathrm{e}^{-rC_{\tau\wedge\overline{y}u}}\mathrm{e}^{-\theta(r)(\tau\wedge\overline{y}u)}] \\
&> \mathbb{E}_{\mathbb{P}}[g_{J_\tau}(r)\mathrm{e}^{-rC_\tau}\mathrm{e}^{-\theta(r)\tau}; \underline{y}u < \tau \le \overline{y}u] \\
&> g_{\min}(r)\mathbb{E}_{\mathbb{P}}[\mathrm{e}^{-\theta(r)\tau} \mid \underline{y}u < \tau \le \overline{y}u]\mathbb{P}[\underline{y}u < \tau \le \overline{y}u] \\
&\ge g_{\min}(r)\mathrm{e}^{-\max\{\underline{y}\theta(r),\overline{y}\theta(r)\}u}\mathbb{P}[\underline{y}u < \tau \le \overline{y}u]\,.
\end{aligned}$$

Hence we obtain the inequality

$$\mathbb{P}[\underline{y}u < \tau \le \overline{y}u] < \frac{\mathbb{E}_{\mathbb{P}}[g_{J_0}(r)]}{g_{\min}(r)}\mathrm{e}^{-\min\{r-\underline{y}\theta(r),r-\overline{y}\theta(r)\}u}\,.$$

Defining the finite time Lundberg exponent

$$R(\underline{y},\overline{y}) = \sup\{\min\{r - \underline{y}\theta(r), r - \overline{y}\theta(r)\} : r \in \mathbb{R}\}$$

we obtain the finite time Lundberg inequality

$$\mathbb{P}[\underline{y}u < \tau \leq \overline{y}u] < \frac{\mathbb{E}_{\mathbb{P}}[g_{J_0}(\tilde{r})]}{g_{\min}(\tilde{r})} \mathrm{e}^{-R(\underline{y},\overline{y})u}$$

where $\tilde{r}$ is the argument maximising $\min\{r - \underline{y}\theta(r), r - \overline{y}\theta(r)\}$. In the discussion of $R(\underline{y}, \overline{y})$ in Sect. 5.14 we only used the property that $\theta(r)$ is convex. Thus the same discussion applies. Let r_y be the argument maximising $r - y\theta(r)$ and let $y_0 = (\theta'(R))^{-1}$ denote the critical value. Then we obtain the two finite time Lundberg inequalities

$$\mathbb{P}[0 < \tau \leq yu \mid C_0 = u] < \frac{\mathbb{E}_{\mathbb{P}}[g_{J_0}(r_y)]}{g_{\min}(r_y)} \mathrm{e}^{-R(0,y)u} ,$$

where $R(0, y) > R$ if $y < y_0$, and

$$\mathbb{P}[yu < \tau < \infty \mid C_0 = u] < \frac{\mathbb{E}_{\mathbb{P}}[g_{J_0}(r_y)]}{g_{\min}(r_y)} \mathrm{e}^{-R(y,\infty)u} ,$$

where $R(y, \infty) > R$ if $y > y_0$.

Let $r_\infty = \sup\{r \in \mathbb{R} : M_Y(r) < \infty\}$. The next result can be proved in the same way as Theorem 5.38.

Theorem 9.9 *Assume that $R < r_\infty$. Then*

$$\frac{\tau}{u} \longrightarrow y_0$$

in probability on the set $\{\tau < \infty\}$. □

Bibliographical Remarks

The Markov modulated risk model was introduced as a semi-Markov model by Janssen [87]. The useful formulation as a model in continuous time goes back to Asmussen [10], where Lemma 9.2, Theorem 9.5 and Theorem 9.7 were also proved. A version of Theorem 9.8, where the limiting constant was dependent on the initial distribution, can be found in [12]. The Theorem presented here is proved in [14]. The model is also discussed in [11, 110].

Appendix A
Stochastic Processes

We start with some definitions.

Definition A.1 Let I be either $\mathbb{N}$ or $[0, \infty)$. A *stochastic process* on I with *state space* E is a family of random variables $\{X_t : t \in I\}$ on E. Let E be a topological space. A stochastic process is called *cadlag* if it has a.s. right continuous paths and the limits from the left exist. It is called *continuous* if its paths are a.s. continuous. We will often identify a cadlag (continuous, respectively) stochastic process with a random variable on the space of cadlag (continuous) functions. In this book we will always assume that stochastic processes on $[0, \infty)$ are cadlag.

Definition A.2 A cadlag stochastic process $\{N_t\}$ on $[0, \infty)$ is called a *point process* if a.s.

(i) $N_0 = 0$,
(ii) $N_t \ge N_s$ for all $t \ge s$ and
(iii) $N_t \in \mathbb{N}$ for all $t \in (0, \infty)$.

We denote the jump times by $T_1, T_2, \ldots$, i.e. $T_k = \inf\{t \ge 0 : N_t \ge k\}$ for all $k \in \mathbb{N}$. In particular, $T_0 = 0$. A point process is called *simple* if $T_0 < T_1 < T_2 < \cdots$.

Definition A.3 A stochastic process $\{X_t\}$ is said to have *independent increments* if for all $n \in \mathbb{N}$ and all real numbers $0 = t_0 < t_1 < t_2 < \cdots < t_n$ the random variables $X_{t_1} - X_{t_0}, X_{t_2} - X_{t_1}, \ldots, X_{t_n} - X_{t_{n-1}}$ are independent. A stochastic process is said to have *stationary increments* if for all $n \in \mathbb{N}$ and all real numbers $0 = t_0 < t_1 < t_2 < \cdots < t_n$ and all $h > 0$ the random vectors $(X_{t_1} - X_{t_0}, X_{t_2} - X_{t_1}, \ldots, X_{t_n} - X_{t_{n-1}})$ and $(X_{t_1+h} - X_{t_0+h}, X_{t_2+h} - X_{t_1+h}, \ldots, X_{t_n+h} - X_{t_{n-1}+h})$ have the same distribution.

Definition A.4 An increasing family $\{\mathcal{F}_t\}$ of σ-algebras is called a *filtration* if $\mathcal{F}_t \subset \mathcal{F}$ for all t. A stochastic process $\{X_t\}$ is called $\{\mathcal{F}_t\}$*-adapted* if X_t is $\mathcal{F}_t$-measurable for all $t \ge 0$. The *natural filtration* $\{\mathcal{F}_t^X\}$ of a stochastic process $\{X_t\}$ is the smallest right continuous filtration such that the process is adapted. A filtration is called *right continuous* if $\mathcal{F}_{t+} := \bigcap_{s>t} \mathcal{F}_s = \mathcal{F}_t$ for all $t \ge 0$. If a stochastic process is considered then, unless stated otherwise, we use its natural filtration.

H. Schmidli, *Risk Theory*, Springer Actuarial,
DOI:10.1007/978-3-319-72005-0

Definition A.5 A $\{\mathcal{F}_t\}$-*stopping time* is a random variable T on $[0, \infty]$ or $\mathbb{N}$ such that $\{T \le t\} \in \mathcal{F}_t$ for all $t \ge 0$. If we stop a stochastic process then we usually assume that T is a stopping time with respect to the natural filtration of the stochastic process. The σ-algebra

$$\mathcal{F}_T := \{A \in \mathcal{F} : A \cap \{T \le t\} \in \mathcal{F}_t \text{ for all } t \ge 0\}$$

is called the information up to the stopping time T.

Bibliographical Remarks

For further literature on stochastic processes see, for instance [42, 58, 89, 90, 111] or [139].

Appendix B
Martingales

Definition B.1 A stochastic process $\{M_t\}$ with state space $\mathbb{R}$ is called an $\{\mathcal{F}_t\}$-*martingale* (-*submartingale*, -*supermartingale*, respectively) if

(i) $\{M_t\}$ is adapted,
(ii) $\mathbb{E}[|M_t|] < \infty$ for all $t \ge 0$,
(iii) $\mathbb{E}[M_t \mid \mathcal{F}_s] = (\ge, \le)\, M_s$ a.s. for all $t \ge s \ge 0$.

We simply say $\{M_t\}$ is a martingale (submartingale, supermartingale) if it is a martingale (submartingale, supermartingale) with respect to its natural filtration.

The following two propositions are very important for dealing with martingales.

Proposition B.2 (Optional stopping theorem) *Let $\{M_t\}$ be an $\{\mathcal{F}_t\}$-martingale (-submartingale, -supermartingale, respectively), and T and S be $\{\mathcal{F}_t\}$-stopping times. Assume that $\{\mathcal{F}_t\}$ is right continuous. Then*

$$\mathbb{E}[M_{S \wedge t} \mid \mathcal{F}_T] = (\ge, \le)\, M_{T \wedge S \wedge t}\ .$$

In particular, the stochastic process $\{M_{T \wedge t} : t \ge 0\}$ is an $\{\mathcal{F}_t\}$-martingale (-submartingale, -supermartingale).

Proof See [42, Thm. 12], [58, Thm. 2.2.13] or [110, Thm. 10.2.5]. □

Proposition B.3 (Martingale convergence theorem) *Let $\{M_t\}$ be an $\{\mathcal{F}_t\}$-submartingale with the property that $\lim_{t \to \infty} \mathbb{E}[M_t^+] < \infty$ (or equivalently $\sup_{t \ge 0} \mathbb{E}[|M_t|] < \infty$). If $\{\mathcal{F}_t\}$ is right continuous then the random variable $M_\infty := \lim_{t \to \infty} M_t$ exists a.s. and is integrable.*

Proof See, for instance [42, Thm. 6] or [110, Thm. 10.2.2]. □

Note that in general for a martingale

H. Schmidli, *Risk Theory*, Springer Actuarial,
DOI:10.1007/978-3-319-72005-0

$$\mathbb{E}[M_\infty] = \mathbb{E}[\lim_{t\to\infty} M_t] \neq \lim_{t\to\infty} \mathbb{E}[M_t] = \mathbb{E}[M_0] .$$

The following result is often useful.

Lemma B.4 (Generalised Borel–Cantelli lemma) *Let* $\{\mathcal{F}_n : n \in \mathbb{N}\}$ *be a filtration,* $A_n \in \mathcal{F}_n$ *and*

$$A_\infty = \cap_n \cup_{m\ge n} A_m = \left\{\sum_{n=1}^{\infty} \mathbb{I}_{A_n} = \infty\right\} .$$

Then

$$\mathbb{P}\left[A_\infty = \left\{\sum_{n=1}^{\infty} \mathbb{P}[A_{n+1} \mid \mathcal{F}_n] = \infty\right\}\right] = 1 .$$

□

Bibliographical Remarks

The theory of martingales can also be found in [42, 58] or [110].

Appendix C
Renewal Processes

C.1 Poisson Processes

Definition C.1 A point process $\{N_t\}$ is called a (homogeneous) *Poisson process* with rate λ if

(i) $\{N_t\}$ has stationary and independent increments,
(ii) $\mathbb{P}[N_h = 0] = 1 - \lambda h + o(h)$ as $h \to 0$,
(iii) $\mathbb{P}[N_h = 1] = \lambda h + o(h)$ as $h \to 0$.

Remark C.2 It follows readily that $\mathbb{P}[N_h \ge 2] = o(h)$ and that the point process is simple.

We now give some alternative definitions of a Poisson process.

Proposition C.3 *Let $\{N_t\}$ be a point process. Then the following are equivalent:*

(i) *$\{N_t\}$ is a Poisson process with rate λ.*
(ii) *$\{N_t\}$ has independent increments and $N_t \sim \mathrm{Pois}(\lambda t)$ for each fixed $t \ge 0$.*
(iii) *The interarrival times $\{T_k - T_{k-1} : k \ge 1\}$ are independent and $\mathrm{Exp}(\lambda)$ distributed.*
(iv) *For each fixed $t \ge 0$, $N_t \sim \mathrm{Pois}(\lambda t)$ and given $\{N_t = n\}$ the occurrence points have the same distribution as the order statistics of n independent uniformly distributed points on $[0, t]$.*
(v) *$\{N_t\}$ has independent increments such that $\mathbb{E}[N_1] = \lambda$ and given $\{N_t = n\}$ the occurrence points have the same distribution as the order statistics of n independent uniformly distributed points on $[0, t]$.*
(vi) *$\{N_t\}$ has independent and stationary increments such that $\mathbb{P}[N_h \ge 2] = o(h)$ and $\mathbb{E}[N_1] = \lambda$.*

Proof "$(i) \Longrightarrow (ii)$" Let $p_n(t) = \mathbb{P}[N_t = n]$. Then show that $p_n(t)$ is continuous and differentiable. Finding the differential equations and solving them shows (ii). The details are left as an exercise.

H. Schmidli, *Risk Theory*, Springer Actuarial,
DOI:10.1007/978-3-319-72005-0

"$(ii) \Longrightarrow (iii)$" We first show that $\{N_t\}$ has stationary increments. It is enough to show that $N_{t+h} - N_h$ is Pois(λt) distributed. For the moment generating functions we obtain

$$e^{\lambda(t+h)(e^r-1)} = \mathbb{E}[e^{r(N_{t+h}-N_h)}e^{rN_h}] = \mathbb{E}[e^{r(N_{t+h}-N_h)}]e^{\lambda h(e^r-1)}\;.$$

It follows that

$$\mathbb{E}[e^{r(N_{t+h}-N_h)}] = e^{\lambda t(e^r-1)}\;.$$

This proves that $N_{t+h} - N_h$ is Pois(λt) distributed.

Let $t_0 = 0 \le s_1 < t_1 \le s_2 < t_2 \le \cdots \le s_n < t_n$. Then

$$\begin{aligned}
&\mathbb{P}[s_k < T_k \le t_k,\ 1 \le k \le n] \\
&\quad = \mathbb{P}[N_{s_k} - N_{t_{k-1}} = 0,\ N_{t_k} - N_{s_k} = 1,\ 1 \le k \le n-1, \\
&\qquad\qquad N_{s_n} - N_{t_{n-1}} = 0,\ N_{t_n} - N_{s_n} \ge 1] \\
&\quad = e^{-\lambda(s_n - t_{n-1})}(1 - e^{-\lambda(t_n - s_n)}) \prod_{k=1}^{n-1} e^{-\lambda(s_k - t_{k-1})}\lambda(t_k - s_k)e^{-\lambda(t_k - s_k)} \\
&\quad = (e^{-\lambda s_n} - e^{-\lambda t_n})\lambda^{n-1} \prod_{k=1}^{n-1}(t_k - s_k) \\
&\quad = \int_{s_1}^{t_1} \cdots \int_{s_n}^{t_n} \lambda^n e^{-\lambda y_n}\, dy_n \cdots dy_1 \\
&\quad = \int_{s_1}^{t_1} \int_{s_2-z_1}^{t_2-z_1} \cdots \int_{s_n-z_1-\cdots-z_{n-1}}^{t_n-z_1-\cdots-z_{n-1}} \lambda^n e^{-\lambda(z_1+\cdots+z_n)}\, dz_n \cdots dz_1\;.
\end{aligned}$$

It follows that the joint density of $T_1, T_2 - T_1, \ldots, T_n - T_{n-1}$ is

$$\lambda^n e^{-\lambda(z_1+\cdots+z_n)}$$

and therefore they are independent and Exp(λ) distributed.

"$(iii) \Longrightarrow (iv)$" Note that T_n is $\Gamma(n, \lambda)$ distributed. Thus $\mathbb{P}[N_t = 0] = \mathbb{P}[T_1 > t] = e^{-\lambda t}$ and for $n \ge 1$

$$\begin{aligned}
\mathbb{P}[N_t = n] &= \mathbb{P}[N_t \ge n] - \mathbb{P}[N_t \ge n+1] = \mathbb{P}[T_n \le t] - \mathbb{P}[T_{n+1} \le t] \\
&= \int_0^t \frac{\lambda^n s^{n-1}}{(n-1)!} e^{-\lambda s}\, ds - \int_0^t \frac{\lambda^{n+1} s^n}{n!} e^{-\lambda s}\, ds \\
&= \int_0^t \frac{d}{ds}\frac{(\lambda s)^n}{n!} e^{-\lambda s}\, ds = \frac{(\lambda t)^n}{n!} e^{-\lambda t}\;.
\end{aligned}$$

The joint density of $T_1, \ldots, T_{n+1}$ is, for $t_0 = 0$,

$$\prod_{k=1}^{n+1} \lambda e^{-\lambda(t_k - t_{k-1})} = \lambda^{n+1} e^{-\lambda t_{n+1}}\,.$$

Thus the joint conditional density of $T_1, \ldots, T_n$ given $N_t = n$ is

$$\frac{\int_t^\infty \lambda^{n+1} e^{-\lambda t_{n+1}} \,\mathrm{d}t_{n+1}}{\int_0^t \int_{t_1}^t \cdots \int_{t_{n-1}}^t \int_t^\infty \lambda^{n+1} e^{-\lambda t_{n+1}} \,\mathrm{d}t_{n+1} \cdots \mathrm{d}t_1} = \frac{n!}{t^n}\,.$$

This is the claimed conditional distribution.

"$(iv) \Longrightarrow (v)$" It is clear that $\mathbb{E}[N_1] = \lambda$. Let $x_k \in \mathbb{N}$ and $t_0 = 0 < t_1 < \cdots < t_n$. Then for $x = x_1 + \cdots + x_n$

$$\begin{aligned}
&\mathbb{P}[N_{t_k} - N_{t_{k-1}} = x_k, 1 \le k \le n] \\
&\quad = \mathbb{P}[N_{t_k} - N_{t_{k-1}} = x_k, 1 \le k \le n \mid N_{t_n} = x]\mathbb{P}[N_{t_n} = x] \\
&\quad = \frac{(\lambda t_n)^x}{x!} e^{-\lambda t_n} \frac{x!}{x_1! \cdots x_n!} \prod_{k=1}^n \Bigl(\frac{t_k - t_{k-1}}{t_n}\Bigr)^{x_k} \\
&\quad = \prod_{k=1}^n \frac{(\lambda(t_k - t_{k-1}))^{x_k}}{x_k!} e^{-\lambda(t_k - t_{k-1})}
\end{aligned}$$

and therefore $\{N_t\}$ has independent increments.

"$(v) \Longrightarrow (i)$" Note that for $x_k \in \mathbb{N}$, $t_0 = 0 < t_1 < \cdots < t_n$ and $h > 0$

$$\begin{aligned}
&\mathbb{P}[N_{t_k+h} - N_{t_{k-1}+h} = x_k, 1 \le k \le n \mid N_{t_n+h}] \\
&\qquad = \mathbb{P}[N_{t_k} - N_{t_{k-1}} = x_k, 1 \le k \le n \mid N_{t_n+h}]\,.
\end{aligned}$$

Integrating with respect to N_{t_n+h} yields that $\{N_t\}$ has stationary increments. For $\mathbb{P}[N_t = 0]$ we obtain

$$\mathbb{P}[N_{t+s} = 0] = \mathbb{P}[N_{t+s} - N_s = 0, N_s = 0] = \mathbb{P}[N_t = 0]\mathbb{P}[N_s = 0]\,.$$

It is left as an exercise to show that $\mathbb{P}[N_t = 0] = (\mathbb{P}[N_1 = 0])^t$. Let now $\lambda_0 := -\log \mathbb{P}[N_1 = 0]$. Then $\mathbb{P}[N_t = 0] = e^{-\lambda_0 t}$. Moreover, for $0 < h < t$,

$$\begin{aligned}
e^{-\lambda_0(t-h)} &= \mathbb{E}[\mathbb{P}[N_{t-h} = 0 \mid N_t]] \\
&= e^{-\lambda_0 t} + \frac{h}{t}\mathbb{P}[N_t = 1] + \sum_{k=2}^\infty \Bigl(\frac{h}{t}\Bigr)^k \mathbb{P}[N_t = k]\,.
\end{aligned}$$

Reordering the terms yields

$$\frac{e^{-\lambda_0(t-h)} - e^{-\lambda_0 t}}{h} t = \mathbb{P}[N_t = 1] + \frac{h}{t} \sum_{k=2}^{\infty} \Big(\frac{h}{t}\Big)^{k-2} \mathbb{P}[N_t = k] \,.$$

Because the sum is bounded by 1, letting $h \to 0$ gives $\mathbb{P}[N_t = 1] = \lambda_0 t e^{-\lambda_0 t}$. In particular, $\{N_t\}$ is a Poisson process with rate λ_0. From "$(i) \Longrightarrow (v)$" we conclude that $\lambda_0 = \mathbb{E}[N_1] = \lambda$. This implies (i).

"$(i) \Longrightarrow (vi)$" Follows readily from (i) and (v).

"$(vi) \Longrightarrow (i)$" As in the proof of "$(v) \Longrightarrow (i)$" it follows that $\mathbb{P}[N_t = 0] = e^{-\lambda_0 t}$. Then $\mathbb{P}[N_h = 1] = 1 - \mathbb{P}[N_h = 0] - \mathbb{P}[N_h \ge 2] = \lambda_0 h + o(h)$. Thus $\{N_t\}$ is a Poisson process with rate λ_0. From "$(i) \Longleftrightarrow (v)$" we conclude that $\lambda_0 = \mathbb{E}[N_1] = \lambda$. □

Remark C.4 The condition $\mathbb{E}[N_1] = \lambda$ in (v) and (vi) is only used for identifying the parameter λ. We did not use in the proof that $\mathbb{E}[N_1] < \infty$. ■

The Poisson process has the following properties.

Proposition C.5 *Let $\{N_t\}$ and $\{\tilde{N}_t\}$ be two independent Poisson processes with rates λ and $\tilde{\lambda}$, respectively. Let $\{I_i : i \in \mathbb{N}\}$ be an iid. sequence of random variables independent of $\{N_t\}$ with $\mathbb{P}[I_i = 1] = 1 - \mathbb{P}[I_i = 0] = q$ for some $q \in (0, 1)$. Furthermore, let $a > 0$ be a real number. Then*

(i) $\{N_t + \tilde{N}_t\}$ *is a Poisson process with rate* $\lambda + \tilde{\lambda}$,

(ii) $\Big\{\sum_{i=1}^{N_t} I_i\Big\}$ *is a Poisson process with rate* λq,

(iii) $\{N_{at}\}$ *is a Poisson process with rate* λa.

Proof Exercise. □

Definition C.6 Let $\Lambda(t)$ be an increasing right continuous function on $[0, \infty)$ with $\Lambda(0) = 0$. A point process $\{N_t\}$ on $[0, \infty)$ is called an *inhomogeneous Poisson process* with *intensity measure* $\Lambda(t)$ if

(i) $\{N_t\}$ has independent increments,
(ii) $N_t - N_s \sim \text{Pois}(\Lambda(t) - \Lambda(s))$.

If there exists a function $\lambda(t)$ such that $\Lambda(t) = \int_0^t \lambda(s)\, ds$ then $\lambda(t)$ is called the *intensity* or *rate* of the inhomogeneous Poisson process.

Note that a homogeneous Poisson process is a special case with $\Lambda(t) = \lambda t$. Define $\Lambda^{-1}(x) = \sup\{t \ge 0 : \Lambda(t) \le x\}$, the inverse function of $\Lambda(t)$.

Proposition C.7 *Let $\{\tilde{N}_t\}$ be a homogeneous Poisson process with rate 1. Define $N_t = \tilde{N}_{\Lambda(t)}$. Then $\{N_t\}$ is an inhomogeneous Poisson process with intensity measure $\Lambda(t)$. Conversely, let $\{N_t\}$ be an inhomogeneous Poisson process with intensity measure $\Lambda(t)$. Let $\tilde{N}_t = N_{\Lambda^{-1}(t)}$ at all points where $\Lambda(\Lambda^{-1}(t)) = t$. On intervals $(\Lambda(\Lambda^{-1}(t)-), \Lambda(\Lambda^{-1}(t)))$ where $\Lambda(\Lambda^{-1}(t)) \neq t$ let there be $N_{\Lambda^{-1}(t)} - N_{(\Lambda^{-1}(t)-)}$ occurrence points uniformly distributed on the interval $(\Lambda(\Lambda^{-1}(t)-), \Lambda(\Lambda^{-1}(t)))$ independent of $\{N_t\}$. Then $\{\tilde{N}_t\}$ is a homogeneous Poisson process with rate* 1.

Proof Exercise. □

For an inhomogeneous Poisson process we can construct the following martingales.

Lemma C.8 *Let $r \in \mathbb{R}$. The following processes are martingales.*

(i) $\{N_t - \Lambda(t)\}$,
(ii) $\{(N_t - \Lambda(t))^2 - \Lambda(t)\}$,
(iii) $\{\exp\{rN_t - \Lambda(t)(e^r - 1)\}\}$.

Proof (i) Because $\{N_t\}$ has independent increments

$$\mathbb{E}[N_t - \Lambda(t) \mid \mathcal{F}_s] = \mathbb{E}[N_t - N_s] + N_s - \Lambda(t) = N_s - \Lambda(s)\,.$$

(ii) Analogously,

$$\begin{aligned}
&\mathbb{E}[(N_t - \Lambda(t))^2 - \Lambda(t) \mid \mathcal{F}_s] \\
&= \mathbb{E}[(N_t - N_s - \{\Lambda(t) - \Lambda(s)\} + N_s - \Lambda(s))^2 \mid \mathcal{F}_s] - \Lambda(t) \\
&= \mathbb{E}[(N_t - N_s - \{\Lambda(t) - \Lambda(s)\})^2] + (N_s - \Lambda(s))^2 \\
&\qquad + 2(N_s - \Lambda(s))\mathbb{E}[N_t - N_s - \{\Lambda(t) - \Lambda(s)\}] - \Lambda(t) \\
&= \Lambda(t) - \Lambda(s) + 2 \cdot 0 + (N_s - \Lambda(s))^2 - \Lambda(t) = (N_s - \Lambda(s))^2 - \Lambda(s)\,.
\end{aligned}$$

(iii) Analogously

$$\mathbb{E}[e^{rN_t - \Lambda(t)(e^r-1)} \mid \mathcal{F}_s] = \mathbb{E}[e^{r(N_t - N_s)}]e^{rN_s - \Lambda(t)(e^r-1)} = e^{rN_s - \Lambda(s)(e^r-1)}\,.$$

□

C.2 Renewal Processes

Definition C.9 A simple point process $\{N_t\}$ is called an *ordinary renewal process* if the interarrival times $\{T_k - T_{k-1} : k \geq 1\}$ are iid. If T_1 has a different distribution then $\{N_t\}$ is called a *delayed renewal process*. If $\lambda^{-1} = \mathbb{E}[T_2 - T_1]$ exists and

$$\mathbb{P}[T_1 \le x] = \lambda \int_0^x \mathbb{P}[T_2 - T_1 > y]\,\mathrm{d}y \tag{C.1}$$

then $\{N_t\}$ is called a *stationary renewal process*. If $\{N_t\}$ is an ordinary renewal process then the function $U(t) = \mathbb{1}_{\{t \ge 0\}} + \mathbb{E}[N_t]$ is called the *renewal measure*.

In the rest of this section we denote by F the distribution function of $T_2 - T_1$. For simplicity we let T be a random variable with distribution F. Note that because the point process is simple we implicitly assume that $F(0) = 0$. Unless stated otherwise we consider in the sequel only ordinary renewal processes.

Recall that F^{*0} is the indicator function of the interval $[0, \infty)$.

Lemma C.10 *The renewal measure can be written as*

$$U(t) = \sum_{n=0}^{\infty} F^{*n}(t)\,.$$

Moreover, $U(t) < \infty$ *for all* $t \ge 0$ *and* $U(t) \to \infty$ *as* $t \to \infty$.

Proof Exercise. □

In renewal theory one often has to solve equations of the form

$$Z(x) = z(x) + \int_0^x Z(x-y)\,\mathrm{d}F(y) \qquad (x \ge 0)\,, \tag{C.2}$$

where $z(x)$ is a known function and $Z(x)$ is unknown. This equation is called the *renewal equation*. The equation can be solved explicitly.

Proposition C.11 *If* $z(x)$ *is bounded on bounded intervals then*

$$Z(x) = \int_{0-}^x z(x-y)\,\mathrm{d}U(y) = z * U(x)$$

is the unique solution to (C.2) *that is bounded on bounded intervals.*

Proof We first show that $Z(x)$ is a solution. This follows from

$$\begin{aligned} Z(x) &= z * U(x) = z * \sum_{n=0}^{\infty} F^{*n}(x) = \sum_{n=0}^{\infty} z * F^{*n}(x) \\ &= z(x) + \sum_{n=1}^{\infty} z * F^{*(n-1)} * F(x) = z(x) + z * U * F(x) \\ &= z(x) + Z * F(x)\,. \end{aligned}$$

Let $Z_1(x)$ be a solution that is bounded on bounded intervals. Then

$$\begin{aligned}|Z_1(x) - Z(x)| &= \left| \int_0^x (Z_1(x-y) - Z(x-y))\, \mathrm{d}F(y) \right| \\ &\leq \int_0^x |Z_1(x-y) - Z(x-y)|\, \mathrm{d}F(y)\end{aligned}$$

and by induction

$$\begin{aligned}|Z_1(x) - Z(x)| &\leq \int_0^x |Z_1(x-y) - Z(x-y)|\, \mathrm{d}F^{*n}(y) \\ &\leq \sup_{0\leq y\leq x} |Z_1(y) - Z(y)| F^{*n}(x)\,.\end{aligned}$$

The latter tends to 0 as $n \to \infty$. Thus $Z_1(x) = Z(x)$. □

Let $z(x)$ be a bounded function and $h > 0$ be a real number. Define

$$\begin{aligned}\overline{m}_k(h) &= \sup\{z(t) : (k-1)h \leq t < kh\}\,, \\ \underline{m}_k(h) &= \inf\{z(t) : (k-1)h \leq t < kh\}\end{aligned}$$

and the Riemann sums

$$\overline{\sigma}(h) = h \sum_{k=1}^{\infty} \overline{m}_k(h), \qquad \underline{\sigma}(h) = h \sum_{k=1}^{\infty} \underline{m}_k(h).$$

Definition C.12 A function $z(x)$ is called *directly Riemann integrable* if

$$-\infty < \liminf_{h\downarrow 0} \underline{\sigma}(h) = \limsup_{h\downarrow 0} \overline{\sigma}(h) < \infty\,.$$

The following lemma gives some criteria for a function to be directly Riemann integrable.

Lemma C.13

- *The space of directly Riemann integrable functions is a linear space.*
- *If $z(t)$ is monotone and $\int_0^\infty z(t)\,\mathrm{d}t < \infty$ then $z(t)$ is directly Riemann integrable.*
- *If $a(t)$ and $b(t)$ are directly Riemann integrable and $z(t)$ is continuous Lebesgue a.e. such that $a(t) \leq z(t) \leq b(t)$ then $z(t)$ is directly Riemann integrable.*
- *If $z(t) \geq 0$, $z(t)$ is continuous Lebesgue a.e. and $\overline{\sigma}(h) < \infty$ for some $h > 0$ then is $z(t)$ directly Riemann integrable.*

Proof See, for instance [1, p. 69]. □

Lemma C.14 *Let $f : [0,\infty) \to [0,\infty)$ be an increasing continuous function and $g : [0,\infty) \to [0,\infty)$ be a decreasing function. If $\int_0^\infty f(x)g(x)\,\mathrm{d}x < \infty$, then $f(x)g(x)$ is directly Riemann integrable.*

Proof We have $\overline{\sigma}(h) \le h\sum_{k=1}^\infty f(kh)g((k-1)h)$. As $h \downarrow 0$, the right-hand side converges to $\int_0^\infty f(x)g(x)\,\mathrm{d}x < \infty$. Further, $\underline{\sigma}(h) \ge h\sum_{k=2}^\infty f((k-1)h)g(kh)$. The right-hand side also converges to $\int_0^\infty f(x)g(x)\,\mathrm{d}x$, which proves the result. □

Definition C.15 A distribution function F of a positive random variable X is called *arithmetic* if for some γ one has $\mathbb{P}[X \in \{\gamma, 2\gamma, \ldots\}] = 1$. The *span* γ is the largest number such that the above equations holds.

The most important result in renewal theory is a result on the asymptotic behaviour of the solution to (C.2).

Proposition C.16 (Key Renewal theorem) *If $z(x)$ is a directly Riemann integrable function then the solution $Z(x)$ to the renewal equation* (C.2) *satisfies*

$$\lim_{t\to\infty} Z(t) = \lambda \int_0^\infty z(y)\,\mathrm{d}y$$

if F is non-arithmetic and

$$\lim_{n\to\infty} Z(t+n\gamma) = \gamma\lambda \sum_{j=0}^\infty z(t+j\gamma)$$

for $0 \le t < \gamma$ if F is arithmetic with span γ.

Proof See, for instance [59, p. 364] or [110, p. 218]. □

Example C.17 Assume that F is non-arithmetic and that $\mathbb{E}[T_1^2] < \infty$. We consider the function $Z(t) = \mathbb{E}[T_{N_t+1} - t]$, the expected time till the next occurrence of an event. Consider first $\mathbb{E}[T_{N_t+1} - t \mid T_1 = s]$. If $s \le t$ then there is a renewal process starting at s and thus $\mathbb{E}[T_{N_t+1} - t \mid T_1 = s] = Z(t-s)$. If $s > t$ then $T_{N_t+1} = T_1 = s$ and thus $\mathbb{E}[T_{N_t+1} - t \mid T_1 = s] = s - t$. Hence we get the renewal equation

$$Z(t) = \mathbb{E}[\mathbb{E}[T_{N_t+1} - t \mid T_1]] = \int_t^\infty (s-t)\,\mathrm{d}F(s) + \int_0^t Z(t-s)\,\mathrm{d}F(s)\,.$$

Now

$$z(t) = \int_t^\infty (s-t)\,\mathrm{d}F(s) = \int_t^\infty \int_t^s \mathrm{d}y\,\mathrm{d}F(s) = \int_t^\infty (1-F(y))\,\mathrm{d}y\,.$$

The function is monotone and

$$\int_0^\infty \int_t^\infty (s-t)\,\mathrm{d}F(s)\,\mathrm{d}t = \int_0^\infty \int_0^s (s-t)\,\mathrm{d}t\,\mathrm{d}F(s) = \frac{1}{2}\mathbb{E}[T_1^2]\,.$$

Thus $z(t)$ is directly Riemann integrable and by the key renewal theorem

$$\lim_{t\to\infty} Z(t) = \frac{\lambda}{2}\mathbb{E}\left[T_1^2\right] .$$

■

Example C.18 Assume that $\lambda > 0$ and that F is non-arithmetic. What is the asymptotic distribution of the waiting time till the next event? Let $x > 0$ and set $Z(t) = \mathbb{P}[T_{N_t+1} - t > x]$. Conditioning on $T_1 = s$ yields

$$\mathbb{P}[T_{N_t+1} - t > x \mid T_1 = s] = \begin{cases} Z(t-s) & \text{if } s \le t, \\ 0 & \text{if } t < s \le t+x, \\ 1 & \text{if } t+x < s. \end{cases}$$

This gives the renewal equation

$$Z(t) = 1 - F(t+x) + \int_0^t Z(t-s)\,\mathrm{d}F(s) .$$

$z(t) = 1 - F(t+x)$ is decreasing in t and its integral is bounded by λ^{-1}. Thus it is directly Riemann integrable. It follows from the key renewal theorem that

$$\lim_{t\to\infty} Z(t) = \lambda \int_0^\infty (1 - F(t+x))\,\mathrm{d}t = \lambda \int_x^\infty (1 - F(t))\,\mathrm{d}t .$$

Thus the stationary distribution of the next event must be the distribution given by (C.1). ■

Bibliographical Remarks

The theory of point processes can also be found in [26] or [39]. For further literature on renewal theory, see also [1, 59] or [110].

Appendix D
Brownian Motion

Definition D.1 A stochastic process $\{W_t\}$ is called an (m, η^2)-*Brownian motion* if a.s.

- $W_0 = 0$,
- $\{W_t\}$ has independent increments,
- $W_t \sim \mathrm{N}(mt, \eta^2 t)$ and
- $\{W_t\}$ has cadlag paths.

A $(0, 1)$-Brownian motion is called a *standard Brownian motion.*

It can be shown that Brownian motion exists. Moreover, one can prove that a Brownian motion has continuous paths.

From the definition it follows that $\{W_t\}$ also has stationary increments.

Lemma D.2 *An (m, η^2)-Brownian motion has stationary increments.*

Proof Exercise. □

We construct the following martingales.

Lemma D.3 *Let $\{W_t\}$ be an (m, η^2)-Brownian motion and $r \in \mathbb{R}$. The following processes are martingales.*

(i) $\{W_t - mt\}$.
(ii) $\{(W_t - mt)^2 - \eta^2 t\}$.
(iii) $\{\exp\{r(W_t - mt) - \frac{\eta^2}{2} r^2 t\}\}$.

Proof (i) Using the stationary and independent increments property

$$\mathbb{E}[W_t - mt \mid \mathcal{F}_s] = \mathbb{E}[W_t - W_s] + W_s - mt = W_s - ms\;.$$

(ii) It suffices to consider the case $m = 0$.

$$\mathbb{E}[W_t^2 - \eta^2 t \mid \mathcal{F}_s] = \mathbb{E}[(W_t - W_s)^2 + 2(W_t - W_s)W_s + W_s^2 \mid \mathcal{F}_s] - \eta^2 t = W_s^2 - \eta^2 s\;.$$

H. Schmidli, *Risk Theory*, Springer Actuarial,
DOI:10.1007/978-3-319-72005-0

(iii) It suffices to consider the case $m = 0$.

$$\mathbb{E}[\mathrm{e}^{rW_t - \frac{\eta^2}{2}r^2 t} \mid \mathcal{F}_s] = \mathbb{E}[\mathrm{e}^{r(W_t - W_s)}]\mathrm{e}^{rW_s - \frac{\eta^2}{2}r^2 t} = \mathrm{e}^{rW_s - \frac{\eta^2}{2}r^2 s}.$$

□

Bibliographical Remarks

More literature on Brownian motion can be found in [58, 89, 110, 111, 139].

Appendix E
Random Walks and the Wiener–Hopf Factorisation

Let $\{X_i\}$ be a sequence of iid. random variables. Let $U(x)$ be the distribution function of X_i. Define

$$S_n = \sum_{i=1}^{n} X_i$$

and $S_0 = 0$.

Lemma E.1
(i) If $\mathbb{E}[X_1] < 0$ then S_n converges to $-\infty$ a.s.
(ii) If $\mathbb{E}[X_1] = 0$ then a.s.

$$\limsup_{n\to\infty} S_n = -\liminf_{n\to\infty} S_n = \infty\;.$$

(iii) If $\mathbb{E}[X_1] > 0$ then S_n converges to ∞ a.s. and there is a strictly positive probability that $S_n \ge 0$ for all $n \in \mathbb{N}$.

Proof See, for instance [59, p. 396] or [110, p. 233]. □

For the rest of this section we assume $-\infty < \mathbb{E}[X_i] < 0$. Let $\tau_+ = \inf\{n > 0 : S_n > 0\}$ and $\tau_- = \inf\{n > 0 : S_n \le 0\}$, $H(x) = \mathbb{P}[\tau_+ < \infty, S_{\tau_+} \le x]$ and $\rho(x) = \mathbb{P}[S_{\tau_-} \le x]$. Note that τ_- is defined a.s. because $S_n \to -\infty$ as $n \to \infty$, see Lemma E.1. Define $\psi_0(x) = \mathbb{I}_{\{x\ge 0\}}$ and

$$\psi_n(x) = \mathbb{P}[S_1 > 0, S_2 > 0, \ldots, S_n > 0, S_n \le x]$$

for $n \ge 1$. Let

$$\psi(x) = \sum_{n=0}^{\infty} \psi_n(x)\;.$$

H. Schmidli, *Risk Theory*, Springer Actuarial,
DOI:10.1007/978-3-319-72005-0

Lemma E.2 *We have for $n \ge 1$*

$$\psi_n(x) = \mathbb{P}[S_n > S_j,\ (0 \le j \le n-1),\ S_n \le x]$$

and therefore

$$\psi(x) = \sum_{n=0}^{\infty} H^{*n}(x)\,.$$

Moreover, $\psi(\infty) = (1 - H(\infty))^{-1}$, $\mathbb{E}[\tau_-] = \psi(\infty)$ and $\mathbb{E}[S_{\tau_-}] = \mathbb{E}[\tau_-]\mathbb{E}[X_i]$.

Proof For n fixed, let

$$S_k^* = S_n - S_{n-k} = \sum_{i=n-k+1}^{n} X_i\,.$$

Then $\{S_k^* : k \le n\}$ follows the same law as $\{S_k : k \le n\}$. Thus

$$\begin{aligned}
&\mathbb{P}[S_n > S_j,\ (0 \le j \le n-1),\ S_n \le x] \\
&\quad = \mathbb{P}[S_n^* > S_j^*,\ (0 \le j \le n-1),\ S_n^* \le x] \\
&\quad = \mathbb{P}[S_n > S_n - S_{n-j},\ (0 \le j \le n-1),\ S_n \le x] \\
&\quad = \mathbb{P}[S_j > 0,\ (1 \le j \le n),\ S_n \le x] = \psi_n(x)\,.
\end{aligned}$$

Denote by $\tau_{n+1} := \inf\{k > 0 : S_k > S_{\tau_n}\}$ the $n+1$-st ascending ladder time. Note that S_{τ_n} has distribution $H^{*n}(x)$. We found that $\psi_n(x)$ is the probability that S_n is a maximum of the random walk and lies in the interval $(0, x]$. Thus $\psi_n(x)$ is the probability that there is a ladder height at n and $S_n \le x$. We obtain

$$\begin{aligned}
\sum_{n=1}^{\infty} \psi_n(x) &= \sum_{n=1}^{\infty} \mathbb{P}[\exists k : \tau_k = n, 0 < S_n \le x] = \sum_{n=1}^{\infty}\sum_{k=1}^{\infty} \mathbb{P}[\tau_k = n, 0 < S_{\tau_k} \le x] \\
&= \sum_{k=1}^{\infty} \mathbb{P}[0 < S_{\tau_k} \le x, \tau_k < \infty] = \sum_{k=1}^{\infty} H^{*k}(x)\,.
\end{aligned}$$

Because $\mathbb{E}[X_i] < 0$ we have $H(\infty) < 1$ (Lemma E.1) and $H^{*n}(\infty) = (H(\infty))^n$ from which $\psi(\infty) = (1 - H(\infty))^{-1}$ follows.

For the expected value of τ_-

$$\mathbb{E}[\tau_-] = \sum_{n=0}^{\infty} \mathbb{P}[\tau_- > n] = \sum_{n=0}^{\infty} \psi_n(\infty) = \psi(\infty)\,.$$

Finally, let $\tau_-(n)$ denote the nth descending ladder time and let L_n be the nth descending ladder height $S_{\tau_-(n-1)} - S_{\tau_-(n)}$. Then

$$\frac{L_1 + \cdots + L_n}{n} = \frac{-S_{\tau_-(n)}}{\tau_-(n)} \frac{\tau_-(n)}{n},$$

from which the last assertion follows by the strong law of large numbers. □

Let now

$$\rho_n(x) = \mathbb{P}[S_1 > 0, S_2 > 0, \ldots, S_{n-1} > 0, S_n \le x].$$

Note that for $x \le 0$

$$\rho(x) = \sum_{n=1}^{\infty} \rho_n(x),$$

that $\psi_1(x) = U(x) - U(0)$ if $x > 0$ and $\rho_1(x) = U(x)$ if $x \le 0$. For $n \ge 1$ one obtains for $x > 0$

$$\psi_{n+1}(x) = \int_0^{\infty} (U(x-y) - U(-y))\psi_n(\mathrm{d}y)$$

and for $x \le 0$

$$\rho_{n+1}(x) = \int_0^{\infty} U(x-y)\psi_n(\mathrm{d}y).$$

If we sum over all n we obtain for $x > 0$

$$\psi(x) - 1 = \int_{0-}^{\infty} (U(x-y) - U(-y))\psi(\mathrm{d}y)$$

and for $x \le 0$

$$\rho(x) = \int_{0-}^{\infty} U(x-y)\psi(\mathrm{d}y).$$

Note that $\rho(0) = \psi(0) = 1$ and thus for $x \ge 0$

$$\psi(x) = \int_{0-}^{\infty} U(x-y)\psi(\mathrm{d}y).$$

The last two equations can be written as a single equation

$$\psi + \rho = \psi_0 + \psi * U.$$

Then

$$\begin{aligned}\psi - \psi_0 + \rho * H &= \psi * H + \rho * H = H + \psi * H * U = H + \psi * U - U \\ &= H - U + \psi + \rho - \psi_0\end{aligned}$$

from which it follows that

$$U = H + \rho - H * \rho\,. \tag{E.1}$$

The latter is called the *Wiener–Hopf factorisation*.

Bibliographical Remarks

The results of this section are taken from [59].

Appendix F
Subexponential Distributions

Definition F.1 A distribution function F with $F(x) = 0$ for $x < 0$ is called *subexponential* if

$$\lim_{t\to\infty} \frac{1 - F^{*2}(t)}{1 - F(t)} = 2\,.$$

The intuition behind the definition is the following. One often observes that the aggregate sum of claims is determined by a few of the largest claims. The definition can be rewritten as $\mathbb{P}[X_1 + X_2 > x] \sim \mathbb{P}[\max\{X_1, X_2\} > x]$. We will prove in Lemma F.7 below that subexponential is equivalent to $\mathbb{P}[X_1 + X_2 + \cdots + X_n > x] \sim \mathbb{P}[\max\{X_1, X_2, \ldots, X_n\} > x]$ for some $n \geq 2$.

We want to show that the moment generating function of such a distribution does not exist for strictly positive values. We first need the following

Lemma F.2 *If F is subexponential then for all $t \in \mathbb{R}$*

$$\lim_{x\to\infty} \frac{1 - F(x-t)}{1 - F(x)} = 1\,.$$

Proof Let $t \geq 0$. We have

$$\begin{aligned}
\frac{1 - F^{*2}(x)}{1 - F(x)} - 1 &= \frac{F(x) - F^{*2}(x)}{1 - F(x)} \\
&= \int_{0-}^{t} \frac{1 - F(x-y)}{1 - F(x)}\,\mathrm{d}F(y) + \int_{t}^{x} \frac{1 - F(x-y)}{1 - F(x)}\,\mathrm{d}F(y) \\
&\geq F(t) + \frac{1 - F(x-t)}{1 - F(x)}(F(x) - F(t))\,.
\end{aligned}$$

Thus

$$1 \leq \frac{1 - F(x-t)}{1 - F(x)} \leq (F(x) - F(t))^{-1}\Bigl(\frac{1 - F^{*2}(x)}{1 - F(x)} - 1 - F(t)\Bigr)\,.$$

H. Schmidli, *Risk Theory*, Springer Actuarial,
DOI:10.1007/978-3-319-72005-0

The assertion for $t \geq 0$ follows by letting $x \to \infty$. If $t < 0$ then

$$\lim_{x\to\infty} \frac{1-F(x-t)}{1-F(x)} = \lim_{x\to\infty} \frac{1}{\dfrac{1-F((x-t)-(-t))}{1-F(x-t)}} = \lim_{y\to\infty} \frac{1}{\dfrac{1-F(y-(-t))}{1-F(y)}} = 1\,,$$

where $y = x - t$. □

Lemma F.3 *Let F be subexponential and $r > 0$. Then*

$$\lim_{t\to\infty} \mathrm{e}^{rt}(1-F(t)) = \infty\,,$$

in particular

$$\int_{0-}^{\infty} \mathrm{e}^{rx}\,\mathrm{d}F(x) = \infty\,.$$

Proof We first observe that for any $0 < x < t$

$$\mathrm{e}^{rt}(1-F(t)) = \frac{1-F(t)}{1-F(t-x)}(1-F(t-x))\mathrm{e}^{r(t-x)}\mathrm{e}^{rx} \tag{F.1}$$

and by Lemma F.2 the function $\mathrm{e}^{rn}(1-F(n))$ $(n \in \mathbb{N})$ is increasing for n large enough. Thus there exists a limit in $(0, \infty]$. Letting $n \to \infty$ in (F.1) shows that this limit can only be infinite.

Let now $\{t_n\}$ be an arbitrary sequence such that $t_n \to \infty$. Then

$$\begin{aligned}\mathrm{e}^{rt_n}(1-F(t_n)) &= \frac{1-F(t_n)}{1-F(\lfloor t_n\rfloor)}(1-F(\lfloor t_n\rfloor))\mathrm{e}^{r\lfloor t_n\rfloor}\mathrm{e}^{r(t_n-\lfloor t_n\rfloor)}\\ &\geq \frac{1-F(t_n)}{1-F(t_n-1)}(1-F(\lfloor t_n\rfloor))e^{\lfloor t_n\rfloor}\,,\end{aligned}$$

which tends to infinity as $n \to \infty$.

The moment generating function can be written as

$$\begin{aligned}\int_{0-}^{\infty} \mathrm{e}^{rx}\,\mathrm{d}F(x) &= 1 + \int_0^{\infty}\int_0^{x} r\mathrm{e}^{ry}\,\mathrm{d}y\,\mathrm{d}F(x)\\ &= 1 + r\int_0^{\infty}\int_y^{\infty} \mathrm{d}F(x)\mathrm{e}^{ry}\,\mathrm{d}y\\ &= 1 + r\int_0^{\infty} \mathrm{e}^{ry}(1-F(y))\,\mathrm{d}y = \infty\end{aligned}$$

because the integrand diverges to infinity. □

The following lemma gives a condition for subexponentiality.

Lemma F.4 *Assume that for all* $z \in (0, 1]$ *the limit*

$$\gamma(z) = \lim_{x\to\infty} \frac{1 - F(zx)}{1 - F(x)}$$

exists and that $\gamma(z)$ *is left continuous at 1. Then* F *is a subexponential distribution function.*

Proof Note first that $F^{*2}(x) = \mathbb{P}[X_1 + X_2 \le x] \le \mathbb{P}[X_1 \le x, X_2 \le x] = F^2(x)$. For simplicity assume $F(0) = 0$. Hence

$$\liminf_{x\to\infty} \frac{1 - F^{*2}(x)}{1 - F(x)} \ge \liminf_{x\to\infty} \frac{1 - F^2(x)}{1 - F(x)} = \liminf_{x\to\infty} 1 + F(x) = 2\,.$$

Let $n \ge 1$ be fixed. Then

$$\begin{aligned}\limsup_{x\to\infty} \frac{1 - F^{*2}(x)}{1 - F(x)} &= 1 + \limsup_{x\to\infty} \int_0^x \frac{1 - F(x-y)}{1 - F(x)}\,\mathrm{d}F(y) \\ &\le 1 + \limsup_{x\to\infty} \sum_{k=1}^{n} \frac{1 - F(x - kx/n)}{1 - F(x)}(F(kx/n) - F((k-1)x/n)) \\ &= 1 + \gamma(1 - 1/n)\,.\end{aligned}$$

Because n is arbitrary and $\gamma(z)$ is left continuous at 1 the assertion follows. □

Example F.5 Consider the Pa(α, β) distribution.

$$\frac{1 - F(zx)}{1 - F(x)} = \frac{(\frac{\beta}{\beta+zx})^\alpha}{(\frac{\beta}{\beta+x})^\alpha} = \Bigl(\frac{\beta + x}{\beta + zx}\Bigr)^\alpha \to z^{-\alpha}$$

as $x \to \infty$. It follows that the Pareto distribution is subexponential. ■

Next we give an upper bound for the tails of the convolutions.

Lemma F.6 *Let* F *be subexponential. Then for any* $\varepsilon > 0$ *there exists a* $D \in \mathbb{R}$ *such that*

$$\frac{1 - F^{*n}(t)}{1 - F(t)} \le D(1 + \varepsilon)^n$$

for all $t > 0$ *and* $n \in \mathbb{N}$.

Proof Let

$$\alpha_n = \sup_{t\ge 0} \frac{1 - F^{*n}(t)}{1 - F(t)}$$

and note that $1 - F^{*(n+1)}(t) = 1 - F(t) + F * (1 - F^{*n})(t)$. Choose $T \geq 0$ such that

$$\sup_{t \geq T} \frac{F(t) - F^{*2}(t)}{1 - F(t)} < 1 + \frac{\varepsilon}{2}.$$

Let $A_T = (1 - F(T))^{-1}$. For $n \geq 1$

$$\begin{aligned}
\alpha_{n+1} &\leq 1 + \sup_{0 \leq t \leq T} \int_{0-}^{t} \frac{1 - F^{*n}(t-y)}{1 - F(t)} \, \mathrm{d}F(y) + \sup_{t \geq T} \int_{0-}^{t} \frac{1 - F^{*n}(t-y)}{1 - F(t)} \, \mathrm{d}F(y) \\
&\leq 1 + A_T + \sup_{t \geq T} \int_{0-}^{t} \frac{1 - F^{*n}(t-y)}{1 - F(t-y)} \frac{1 - F(t-y)}{1 - F(t)} \, \mathrm{d}F(y) \\
&\leq 1 + A_T + \alpha_n \sup_{t \geq T} \frac{F(t) - F^{*2}(t)}{1 - F(t)} \leq 1 + A_T + \alpha_n \Big(1 + \frac{\varepsilon}{2}\Big).
\end{aligned}$$

Choose

$$D = \max\Big\{\frac{2(1 + A_T)}{\varepsilon}, 1\Big\}$$

and note that $\alpha_1 = 1 < D(1 + \varepsilon)$. The assertion follows by induction. □

Lemma F.7 *Let $F(x) = 0$ for $x < 0$. The following are equivalent:*

(i) F is subexponential.
(ii) For all $n \in \mathbb{N}$

$$\lim_{x \to \infty} \frac{1 - F^{*n}(x)}{1 - F(x)} = n. \tag{F.2}$$

(iii) There exists an $n \geq 2$ such that (F.2) *holds.*

Proof "$(i) \Longrightarrow (ii)$" We prove the assertion by induction. The assertion is trivial for $n \leq 2$. Assume that it is proved for some $n \in \mathbb{N}$ with $n \geq 2$. Let $\varepsilon > 0$. Choose t such that

$$\Big|\frac{1 - F^{*n}(x)}{1 - F(x)} - n\Big| < \varepsilon$$

for $x \geq t$.

$$\begin{aligned}
\frac{1 - F^{*(n+1)}(x)}{1 - F(x)} &= 1 + \int_{0-}^{x-t} \frac{1 - F^{*n}(x-y)}{1 - F(x-y)} \frac{1 - F(x-y)}{1 - F(x)} \, \mathrm{d}F(y) \\
&\quad + \int_{x-t}^{x} \frac{1 - F^{*n}(x-y)}{1 - F(x)} \, \mathrm{d}F(y).
\end{aligned}$$

The second integral is bounded by

$$\int_{x-t}^{x} \frac{1 - F^{*n}(x-y)}{1 - F(x)} \, \mathrm{d}F(y) \leq \frac{F(x) - F(x-t)}{1 - F(x)} = \frac{1 - F(x-t)}{1 - F(x)} - 1$$

which tends to 0 as $x \to \infty$ by Lemma F.2. The expression

$$\int_{0-}^{x-t} n \frac{1-F(x-y)}{1-F(x)} \,\mathrm{d}F(y) \\ = n\left(\frac{F(x)-F^{*2}(x)}{1-F(x)} - \int_{x-t}^{x} \frac{1-F(x-y)}{1-F(x)} \,\mathrm{d}F(y)\right) \to n$$

as $x \to \infty$ by the same arguments as above. Then

$$\left|\int_{0-}^{x-t} \left(\frac{1-F^{*n}(x-y)}{1-F(x-y)} - n\right) \frac{1-F(x-y)}{1-F(x)} \,\mathrm{d}F(y)\right| \\ \le \varepsilon\left(\frac{F(x)-F^{*2}(x)}{1-F(x)} - \int_{x-t}^{x} \frac{1-F(x-y)}{1-F(x)} \,\mathrm{d}F(y)\right) \to \varepsilon$$

by the arguments used before. Thus

$$\limsup_{x\to\infty}\left|\frac{1-F^{*(n+1)}(x)}{1-F(x)} - (n+1)\right| \le \varepsilon\,.$$

This proves the assertion because ε was arbitrary.

"$(ii) \Longrightarrow (iii)$" Trivial.

"$(iii) \Longrightarrow (i)$" Assume now (F.2) for some $n > 2$. We show that the assertion also holds for n replaced by $n-1$. Since

$$\frac{1-F^{*n}(x)}{1-F(x)} - 1 = \int_{0-}^{x} \frac{1-F^{*(n-1)}(x-y)}{1-F(x)} \,\mathrm{d}F(y) \ge F(x)\frac{1-F^{*(n-1)}(x)}{1-F(x)}$$

it follows that

$$\limsup_{x\to\infty} \frac{1-F^{*(n-1)}(x)}{1-F(x)} \le n-1\,.$$

Because $F^{*(n-1)}(x) \le F(x)^{n-1}$ it is trivial that

$$\liminf_{x\to\infty} \frac{1-F^{*(n-1)}(x)}{1-F(x)} \ge n-1\,.$$

It follows that (F.2) holds for $n = 2$ and F is a subexponential distribution. □

Lemma F.8 *Let U and V be two distribution functions with $U(x) = V(x) = 0$ for all $x < 0$. Assume that*

$$1 - V(x) \sim a(1-U(x)) \qquad \text{as } x \to \infty$$

for some $a > 0$. If U is subexponential then V is also subexponential.

Proof We have to show that

$$\limsup_{x\to\infty} \int_{0-}^{x} \frac{1 - V(x-y)}{1 - V(x)} \, \mathrm{d}V(y) \le 1 \,.$$

Let $0 < \varepsilon < 1$. There exists a y_0 such that for all $y \ge y_0$

$$(1-\varepsilon)a \le \frac{1 - V(y)}{1 - U(y)} \le (1+\varepsilon)a \,.$$

Let $x > y_0$. Note that

$$\begin{aligned}\int_{x-y_0}^{x} \frac{1 - V(x-y)}{1 - V(x)} \, \mathrm{d}V(y) &\le \frac{V(x) - V(x-y_0)}{1 - V(x)} \\ &= \frac{\frac{1-V(x-y_0)}{1-U(x-y_0)}}{\frac{1-V(x)}{1-U(x)}} \frac{1 - U(x-y_0)}{1 - U(x)} - 1 \to 0\end{aligned}$$

by Lemma F.2. Moreover,

$$\begin{aligned}\int_{0-}^{x-y_0} \frac{1 - V(x-y)}{1 - V(x)} \, \mathrm{d}V(y) &\le \frac{1+\varepsilon}{1-\varepsilon} \int_{0-}^{x-y_0} \frac{1 - U(x-y)}{1 - U(x)} \, \mathrm{d}V(y) \\ &\le \frac{1+\varepsilon}{1-\varepsilon} \int_{0-}^{x} \frac{1 - U(x-y)}{1 - U(x)} \, \mathrm{d}V(y) \,.\end{aligned}$$

The last integral is

$$\begin{aligned}\int_{0-}^{x} \frac{1 - U(x-y)}{1 - U(x)} \, \mathrm{d}V(y) &= \frac{V(x) - U * V(x)}{1 - U(x)} \\ &= 1 - \frac{1 - V(x)}{1 - U(x)} + \frac{U(x) - V * U(x)}{1 - U(x)} \\ &= 1 - \frac{1 - V(x)}{1 - U(x)} + \int_{0-}^{x} \frac{1 - V(x-y)}{1 - U(x)} \, \mathrm{d}U(y) \,.\end{aligned}$$

$1 - (1 - U(x))^{-1}(1 - V(x))$ tends to $1 - a$ as $x \to \infty$ and

$$\int_{x-y_0}^{x} \frac{1 - V(x-y)}{1 - U(x)} \, \mathrm{d}U(y) \le \frac{U(x) - U(x-y_0)}{1 - U(x)}$$

tends to 0. Finally,

$$\int_{0-}^{x-y_0} \frac{1-V(x-y)}{1-U(x)}\,\mathrm{d}U(y) \le (1+\varepsilon)a \int_{0-}^{x-y_0} \frac{1-U(x-y)}{1-U(x)}\,\mathrm{d}U(y)$$
$$\le (1+\varepsilon)a \int_{0-}^{x} \frac{1-U(x-y)}{1-U(x)}\,\mathrm{d}U(y)\,,$$

which tends to $(1+\varepsilon)a$. Putting all these limits together we obtain

$$\limsup_{x\to\infty} \int_{0-}^{x} \frac{1-V(x-y)}{1-V(x)}\,\mathrm{d}V(y) \le \frac{(1+\varepsilon)(1+a\varepsilon)}{1-\varepsilon}\,.$$

Because ε was arbitrary the assertion follows. □

In applications we often need that $G^{\mathrm{s}} = \mu^{-1}\int_0^x (1-G(y))\,\mathrm{d}y$ is subexponential, see Theorem 5.26 or Theorem 6.14. In order to show that G^{s} is subexponential it has to satisfy

$$\lim_{x\to\infty} \frac{\int_x^\infty \int_y^\infty (1-G(y-z))(1-G(z))\,\mathrm{d}z\,\mathrm{d}y}{\mu\int_x^\infty (1-G(z))\,\mathrm{d}z} = 2\,.$$

Applying L'Hospital's rule gives

$$\lim_{x\to\infty} \frac{\int_x^\infty (1-G(x-z))(1-G(z))\,\mathrm{d}z\,\mathrm{d}y}{\mu(1-G(x))} = 2\,.$$

Definition F.9 We say a distribution function $F(x)$ with $F(x)=0$ for $x<0$ is in the class $\mathcal{S}^*$ if

$$\lim_{x\to\infty} \int_x^\infty \frac{(1-F(x-z))(1-F(z))}{1-F(x)}\,\mathrm{d}z = 2\mu_F\,,$$

where $\mu_F = \int_0^\infty (1-F(z))\,\mathrm{d}z$.

By construction, if $F\in\mathcal{S}^*$, then F^{s} is subexponential.

Proposition F.10 *If $F\in\mathcal{S}^*$, then F and F^{s} are subexponential.*

Proof See [92]. □

If F has density $F'(x)$, we can define the hazard rate $\ell_F(x) = F'(x)/(1-F(x))$. The following criteria are helpful to verify that $F\in\mathcal{S}^*$.

Lemma F.11 *Let F be a distribution function with hazard rate $\ell_F(x)$ and $F(0)=0$.*

(i) If G and H are distribution functions with $G(x)=H(x)=0$ for $x<0$ and $\lim_{x\to\infty}(1-G(x))/(1-H(x)) = c\in(0,\infty)$, then $G\in\mathcal{S}^*$ if and only if $H\in\mathcal{S}^*$.
(ii) If $\mu_F<\infty$ and $\limsup_{x\to\infty} x\ell_F(x)<\infty$, then $F\in\mathcal{S}^*$.

(iii) If $\ell_F(x)$ is ultimately decreasing, and $\lim_{x\to\infty} \ell_F(x) = 0$ *such that*

$$\int_0^\infty \exp\{x\ell_F(x)\}(1 - F(x))\, \mathrm{d}x < \infty \,,$$

then $F \in \mathcal{S}^*$.

Proof (i) can be verified similarly to Lemma F.8. The other results are proved in [92]. □

From (ii) and (i), it follows from Karamata's theorem that if $1 - F(x)$ is regularly varying, then $F \in \mathcal{F}^*$. (iii) gives that the Weibull distribution with $\alpha \in (0, 1)$ and the lognormal distribution are in $\mathcal{S}^*$.

Bibliographical Remarks

The class of subexponential distributions was introduced by Chistyakov [35]. The results presented here can be found in [15, 35, 54]. The class $\mathcal{S}^*$ was introduced in [92], where the results on S^* can be found.

Appendix G
Concave and Convex Functions

In this appendix we let I be an interval, finite or infinite, but not a singleton.

Definition G.1 A function $u : I \to \mathbb{R}$ is called *(strictly) concave* if for all $x, z \in I$, $x \neq z$, and all $\alpha \in (0, 1)$ one has

$$u((1-\alpha)x + \alpha z) \geq (>)\,(1-\alpha)u(x) + \alpha u(z)\,.$$

u is called *(strictly) convex* if $-u$ is (strictly) concave.

Because results on concave functions can easily translated to convex functions we will only consider concave functions in the sequel.

Concave functions have nice properties.

Lemma G.2 *A concave function $u(y)$ is continuous and differentiable from the left and from the right. The derivative is decreasing, i.e. for $x < y$ we have $u'(x-) \geq u'(x+) \geq u'(y-) \geq u'(y+)$. If $u(y)$ is strictly concave then $u'(x+) > u'(y-)$.*

Remark G.3 The lemma implies that $u(y)$ is differentiable almost everywhere. ■

Proof Let $x < y < z$. Then

$$u(y) = u\Bigl(\frac{z-y}{z-x}x + \frac{y-x}{z-x}z\Bigr) \geq \frac{z-y}{z-x}u(x) + \frac{y-x}{z-x}u(z)$$

or equivalently

$$(z-x)u(y) \geq (z-y)u(x) + (y-x)u(z)\,. \tag{G.1}$$

This implies immediately

$$\frac{u(y)-u(x)}{y-x} \geq \frac{u(z)-u(x)}{z-x} \geq \frac{u(z)-u(y)}{z-y}\,. \tag{G.2}$$

Thus the function $h \mapsto h^{-1}(u(y) - u(y-h))$ is increasing in h and bounded from below by $(z-y)^{-1}(u(z) - u(y))$. Thus the derivative $u'(y-)$ from the left exists.

H. Schmidli, *Risk Theory*, Springer Actuarial,
DOI:10.1007/978-3-319-72005-0

Analogously, the derivative from the right $u'(y+)$ exists. The assertion in the concave case now follows from (G.2). The strict inequality in the strictly concave case follows analogously. □

Concave functions also have the following property.

Lemma G.4 *Let $u(y)$ be a concave function. There exists a function $k : I \to \mathbb{R}$ such that for any $y, x \in I$*

$$u(x) \le u(y) + k(y)(x - y) . \tag{G.3}$$

Moreover, the function $k(y)$ is decreasing. If $u(y)$ is strictly concave then the above inequality is strict for $x \ne y$ and $k(y)$ is strictly decreasing. Conversely, if a function $k(y)$ exists such that (G.3) *is satisfied, then $u(y)$ is concave, strictly concave if the strict inequality holds for $x \ne y$.*

Proof Left as an exercise. □

Corollary G.5 *Let $u(y)$ be a twice differentiable function. Then $u(y)$ is concave if and only if its second derivative is negative. It is strictly concave if and only if its second derivative is strictly negative almost everywhere.*

Proof This follows readily from Lemmata G.2 and G.4. □

The following result is very useful.

Theorem G.6 (Jensen's inequality) *The function $u(y)$ is (strictly) concave if and only if*

$$\mathbb{E}[u(Y)] \le (<)\, u(\mathbb{E}[Y]) \tag{G.4}$$

for all I-valued integrable random variables Y with $\mathbb{P}[Y \ne \mathbb{E}[Y]] > 0$.

Proof Assume (G.4) for all random variables Y. Let $\alpha \in (0, 1)$. Let $\mathbb{P}[Y = z] = 1 - \mathbb{P}[Y = x] = \alpha$. Then the (strict) concavity follows. Assume $u(y)$ is strictly concave. Then it follows from Lemma G.4 that

$$u(Y) \le u(\mathbb{E}[Y]) + k(\mathbb{E}[Y])(Y - \mathbb{E}[Y]) .$$

The strict inequality holds if $u(y)$ is strictly concave and $Y \ne \mathbb{E}[Y]$. Taking expected values gives (G.4). □

The following result is also often useful.

Theorem G.7 (Ohlin's lemma) *Let $F_i(y)$, $i = 1, 2$, be two distribution functions defined on I. Assume*

$$\int_I y \,\mathrm{d}F_1(y) = \int_I y \,\mathrm{d}F_2(y) < \infty$$

and that there exists a $y_0 \in I$ *such that*

$$F_1(y) \leq F_2(y), \quad y < y_0, \qquad F_1(y) \geq F_2(y), \quad y > y_0 \,.$$

Then for any concave function $u(y)$

$$\int_I u(y)\,\mathrm{d}F_1(y) \geq \int_I u(y)\,\mathrm{d}F_2(y)$$

provided the integrals are well defined. If $u(y)$ *is strictly concave and* $F_1 \neq F_2$ *then the inequality holds strictly.*

Proof Recall the formulae

$$\int_0^\infty y\,\mathrm{d}F_i(y) = \int_0^\infty (1 - F_i(y))\,\mathrm{d}y \quad \text{and} \quad \int_{-\infty}^0 y\,\mathrm{d}F_i(y) = -\int_{-\infty}^0 F_i(y)\,\mathrm{d}y \,,$$

which can be proved, for example, by Fubini's theorem. Thus it follows that $\int_I (F_2(y) - F_1(y))\,\mathrm{d}y = 0$. We know that $u(y)$ is differentiable almost everywhere and continuous. Thus $u(y) = u(y_0) + \int_{y_0}^y u'(z)\,\mathrm{d}z$, where we can define $u'(y)$ as the derivative from the right. This yields

$$\begin{aligned}\int_{-\infty}^{y_0} u(y)\,\mathrm{d}F_i(y) &= u(y_0)F_i(y_0) - \int_{-\infty}^{y_0}\int_y^{y_0} u'(z)\,\mathrm{d}z\,\mathrm{d}F_i(y) \\ &= u(y_0)F_i(y_0) - \int_{-\infty}^{y_0} F_i(z)u'(z)\,\mathrm{d}z \,.\end{aligned}$$

Analogously

$$\int_{y_0}^\infty u(y)\,\mathrm{d}F_i(y) = u(y_0)(1 - F_i(y_0)) + \int_{y_0}^\infty (1 - F_i(z))u'(z)\,\mathrm{d}z \,.$$

Putting the results together we find

$$\int_{-\infty}^\infty u(y)\,\mathrm{d}F_1(y) - \int_{-\infty}^\infty u(y)\,\mathrm{d}F_2(y) = \int_{-\infty}^\infty (F_2(y) - F_1(y))u'(y)\,\mathrm{d}y \,.$$

If $y < y_0$ then $F_2(y) - F_1(y) \geq 0$ and $u'(y) \geq u'(y_0)$. If $y > y_0$ then $F_2(y) - F_1(y) \leq 0$ and $u'(y) \leq u'(y_0)$. Thus

$$\int_{-\infty}^\infty u(y)\,\mathrm{d}F_1(y) - \int_{-\infty}^\infty u(y)\,\mathrm{d}F_2(y) \geq \int_{-\infty}^\infty (F_2(y) - F_1(y))u'(y_0)\,\mathrm{d}y = 0 \,.$$

The strictly concave case follows analogously. □

Corollary G.8 *Let X be a real random variable taking values in some interval I_1, and let $g_i : I_1 \to I_2$, $i = 1, 2$ be increasing functions with values on some interval I_2. Suppose*

$$\mathbb{E}[g_1(X)] = \mathbb{E}[g_2(X)] < \infty .$$

Let $u : I_2 \to \mathbb{R}$ be a concave function such that $\mathbb{E}[u(g_i(X))]$ is well-defined. If there exists an x_0 such that

$$g_1(x) \geq g_2(x), \quad x < x_0, \qquad\qquad g_1(x) \leq g_2(x), \quad x > x_0,$$

then

$$\mathbb{E}[u(g_1(X))] \geq \mathbb{E}[u(g_2(X))] .$$

Moreover, if $u(y)$ is strictly concave and $\mathbb{P}[g_1(X) \neq g_2(X)] > 0$ then the inequality is strict.

Proof Choose $F_i(y) = \mathbb{P}[g_i(X) \leq y]$. Let $y_0 = g_1(x_0)$. If $y < y_0$ then

$$\begin{aligned} F_1(y) = \mathbb{P}[g_1(X) \leq y] &= \mathbb{P}[g_1(X) \leq y, X < x_0] \\ &\leq \mathbb{P}[g_2(X) \leq y, X < x_0] \leq F_2(y) . \end{aligned}$$

If $y > y_0$ then

$$1 - F_1(y) = \mathbb{P}[g_1(X) > y, X > x_0] \leq \mathbb{P}[g_2(X) > y, X > x_0] \leq 1 - F_2(y) .$$

The result now follows from Theorem G.7. □

Bibliographical Remarks

Ohlin's lemma and the corollary can be found in [105].

Table of Distribution Functions

Exponential distribution: Exp(α)

Parameters:	$\alpha > 0$
Distribution function:	$F(x) = 1 - \mathrm{e}^{-\alpha x}$
Density:	$f(x) = \alpha \mathrm{e}^{-\alpha x}$
Expected value:	$\mathbb{E}[X] = \alpha^{-1}$
Variance:	$\mathrm{Var}(X) = \alpha^{-2}$
Moment generating fnc:	$M_X(r) = \alpha(\alpha - r)^{-1}$

Gamma distribution: $\Gamma(\gamma, \alpha)$

Parameters:	$\gamma > 0,\ \alpha > 0$
Distribution function:	$F(x)$= — —
Density:	$f(x) = \alpha^{\gamma}(\Gamma(\gamma))^{-1} x^{\gamma-1} \mathrm{e}^{-\alpha x}$
Expected value:	$\mathbb{E}[X] = \gamma\alpha^{-1}$
Variance:	$\mathrm{Var}(X) = \gamma\alpha^{-2}$
Moment generating fnc:	$M_X(r) = (\alpha(\alpha - r)^{-1})^{\gamma}$

Weibull distribution: Wei(α, c)

Parameters:	$\alpha > 0, c > 0$
Distribution function:	$F(x) = 1 - \mathrm{e}^{-cx^{\alpha}}$
Density:	$f(x) = \alpha c x^{\alpha-1} \mathrm{e}^{-cx^{\alpha}}$
Expected value:	$\mathbb{E}[X] = c^{-1/\alpha}\Gamma(1 + \frac{1}{\alpha})$
Variance:	$\mathrm{Var}(X) = c^{-2/\alpha}(\Gamma(1 + \frac{2}{\alpha}) - \Gamma(1 + \frac{1}{\alpha})^2)$
Moment generating fnc:	$M_X(r)$ = — —

H. Schmidli, *Risk Theory*, Springer Actuarial,
DOI:10.1007/978-3-319-72005-0

Pareto distribution: Pa(α, β)

Parameters:	$\alpha > 0, \beta > 0$
Distribution function:	$F(x) = 1 - \beta^{\alpha}(\beta + x)^{-\alpha}$
Density:	$f(x) = \alpha\beta^{\alpha}(\beta + x)^{-\alpha-1}$
Expected value:	$\mathbb{E}[X] = \beta(\alpha - 1)^{-1}$ if $\alpha > 1$
Variance:	$\mathrm{Var}(X) = \alpha\beta^2(\alpha - 1)^{-2}(\alpha - 2)^{-1}$ if $\alpha > 2$
Moment generating fnc:	$M_X(r) =$ — —

Loggamma distribution: LG(γ, α)

Parameters:	$\gamma > 0, \ \alpha > 0$
Distribution function:	$F(x) =$ — —
Density:	$f(x) = \alpha^{\gamma}(\Gamma(\gamma))^{-1}x^{-\alpha-1}(\log x)^{\gamma-1} \quad (x > 1)$
Expected value:	$\mathbb{E}[X] = (\alpha(\alpha - 1)^{-1})^{\gamma}$ if $\alpha > 1$
Variance:	$\mathrm{Var}(X) = (\alpha/(\alpha - 2))^{\gamma} - (\alpha/(\alpha - 1))^{2\gamma}, \alpha > 2$
Moment generating fnc:	$M_X(r) =$ — —

Beta distribution: $B(\alpha, \beta)$

Parameters:	$\alpha, \beta > 0$
Distribution function:	$F(x) =$ — —
Density:	$f(x) = \frac{\Gamma(\alpha+\beta)}{\Gamma(\alpha)\Gamma(\beta)}x^{\alpha-1}(1 - x)^{\beta-1}\mathbb{1}_{x\in(0,1)}$
Expected value:	$\mathbb{E}[X] = (\alpha + \beta)^{-1}\alpha$
Variance:	$\mathrm{Var}(X) = ((\alpha + \beta)^2(\alpha + \beta + 1))^{-1}\alpha\beta$
Moment generating fnc:	$M_X(r) =$ — —

Normal distribution: N(μ, σ^2)

Parameters:	$\mu \in \mathbb{R}, \ \sigma^2 > 0$
Distribution function:	$F(x) = \Phi(\sigma^{-1}(x - \mu))$
Density:	$f(x) = (2\pi\sigma^2)^{-1/2}\mathrm{e}^{-(x-\mu)^2/(2\sigma^2)}$
Expected value:	$\mathbb{E}[X] = \mu$
Variance:	$\mathrm{Var}(X) = \sigma^2$
Moment generating fnc:	$M_X(r) = \exp\{(\sigma^2/2)r^2 + r\mu\}$

Lognormal distribution: LN(μ, σ^2)

Parameters:	$\mu \in \mathbb{R}, \ \sigma^2 > 0$
Distribution function:	$F(x) = \Phi(\sigma^{-1}(\log x - \mu))$
Density:	$f(x) = (2\pi\sigma^2)^{-1/2}x^{-1}\mathrm{e}^{-(\log x-\mu)^2/(2\sigma^2)}$
Expected value:	$\mathbb{E}[X] = \mathrm{e}^{(\sigma^2/2)+\mu}$
Variance:	$\mathrm{Var}(X) = \mathrm{e}^{\sigma^2+2\mu}(\mathrm{e}^{\sigma^2} - 1)$
Moment generating fnc:	$M_X(r) =$ — —

Binomial distribution: B(k, p)

Parameters:	$k \in \mathbb{N}, \ p \in (0, 1)$
Probabilities:	$\mathbb{P}[X = n] = \binom{k}{n}p^n(1 - p)^{k-n}$
Expected value:	$\mathbb{E}[X] = kp$
Variance:	$\mathrm{Var}(X) = kp(1 - p)$
Moment generating function:	$M_X(r) = (p\mathrm{e}^r + 1 - p)^k$

Poisson distribution: Pois(λ)

Parameters:	$\lambda > 0$
Probabilities:	$\mathbb{P}[X = n] = (n!)^{-1}\lambda^n e^{-\lambda}$
Expected value:	$\mathbb{E}[X] = \lambda$
Variance:	$\text{Var}(X) = \lambda$
Moment generating function:	$M_X(r) = \exp\{\lambda(e^r - 1)\}$

Negative binomial distribution: NB(α, p)

Parameters:	$\alpha > 0,\ p \in (0, 1)$
Probabilities:	$\mathbb{P}[X = n] = \binom{\alpha+n-1}{n} p^\alpha (1-p)^n$
Expected value:	$\mathbb{E}[X] = \alpha(1-p)/p$
Variance:	$\text{Var}(X) = \alpha(1-p)/p^2$
Moment generating function:	$M_X(r) = p^\alpha (1 - (1-p)e^r)^{-\alpha}$

References

1. Alsmeyer G (1991) Erneuerungstheory. Teubner, Stuttgart
2. **Ammeter, H.** (1948). A generalization of the collective theory of risk in regard to fluctuating basic probabilities. *Skand. Aktuar Tidskr.*, 171–198
3. Andersen E Sparre (1957) On the collective theory of risk in the case of contagion between the claims. Transactions XVth International Congress of Actuaries, New York II:219–229
4. **Arfwedson, G.** (1954). Research in collective risk theory. Part 1. *Skand. Aktuar Tidskr.*, 191–223
5. **Arfwedson, G.** (1955). Research in collective risk theory. Part 2. *Skand. Aktuar Tidskr.*, 53–100
6. Arrow KJ (1965) Aspects of the Theory of Risk Bearing. Yrjo Jahnssonin Saatio, Helsinki
7. Artzner P, Delbaen F, Eber J-M, Heath D (1999) Coherent measures of risk. Math. Finance 9:203–228
8. Asmussen S (1982) Conditioned limit theorems relating a random walk to its associate, with applications to risk reserve processes and the GI/G/1 Queue. Adv. in Appl. Probab. 14:143–170
9. **Asmussen, S.** (1984). Approximations for the probability of ruin within finite time. *Scand. Actuarial J.*, 31–57
10. **Asmussen, S.** (1989). Risk theory in a Markovian environment. *Scand. Actuarial J.*, 66–100
11. Asmussen S, Albrecher H (2010) Ruin Probabilities, 2nd edn. World Scientific, Singapore
12. Asmussen S, Fløe Henriksen F, Klüppelberg C (1994) Large claims approximations for risk processes in a Markovian environment. Stochastic Process. Appl. 54:29–43
13. **Asmussen, S.** and **Nielsen, H.M.** (1995). Ruin probabilities via local adjustment coefficients. *J. Appl. Probab.*, 736–755
14. Asmussen S, Schmidli H, Schmidt V (1999) Tail probabilities for non-standard risk and queueing processes with subexponential jumps. Adv. in Appl. Probab. 31:422–447
15. Athreya KB, Ney P (2004) Branching Processes. Dover Publications Inc, Mineola, NY
16. Bailey AL (1945) A generalised theory of credibility. Proceedings of the Casualty Actuarial Society 32:13–20
17. **Bailey, A.L.** (1950). Credibility procedures, La Place's generalisation of Bayes' rule, and the combination of collateral knowledge with observed data. *Proceedings of the Casualty Actuarial Society* **37**, 7–23 and 94–115
18. **Barndorff-Nielsen, O.E.** and **Schmidli, H.** (1994). Saddlepoint approximations for the probability of ruin in finite time. *Scand. Actuarial J.*, 169–186
19. Beard RE, Pentikainen T, Pesonen E (1984) Risk Theory, 3rd edn. Chapman & Hall Ltd., London

H. Schmidli, *Risk Theory*, Springer Actuarial,
DOI:10.1007/978-3-319-72005-0

20. **Beekman, J.** (1969). A ruin function approximation. *Trans. Soc. Actuaries***21**, 41–48 and 275–279
21. Bertoin J (1996) Lévy processes. Cambridge University Press, Cambridge
22. **Björk, T.** and **Grandell, J.** (1988). Exponential inequalities for ruin probabilities in the Cox case. *Scand. Actuarial J.*, 77–111
23. Borch K (1962) Equilibrium in a reinsurance market. Econometrica 30:424–444
24. Borch K (1975) Optimal insurance arrangments. ASTIN Bull. 8:284–290
25. Bornhuetter RL, Ferguson RE (1972) The actuary and IBNR. Proc. CAS 59:181–195
26. Brémaud P (1981) Point Processes and Queues. Springer-Verlag, New York
27. Bühlmann H (1967) Experience rating and credibility. ASTIN Bull. 4:199–207
28. Bühlmann H (1970) Mathematical Methods in Risk Theory. Springer-Verlag, New York
29. Bühlmann H (1987) Premium calculation from top down. ASTIN Bull. 15:89–102
30. Bühlmann H, Jewell WS (1979) Optimal risk exchanges. ASTIN Bull. 10:243–263
31. Bühlmann H, Straub E (1970) Glaubwürdigkeit für Schadensätze. Schweiz. Verein. Versicherungsmath. Mitt. 70:111–133
32. Cass RM, Kensicki PR, Patrik GS, Reinarz RC (1997) Reinsurance Practices, 2nd edn. Insuance Institute of America, Malvern, PA
33. Centeno L (1986) Measuring the effects of reinsurance by the adjustment coefficient. Insurance Math. Econom. 5:169–182
34. Cheridito P, Delbaen F, Kupper M (2006) Coherent and convex monetary risk measures for unbounded càdlàg processes. Finance Stoch. 10:427–448
35. Chistyakov VP (1964) A theorem on sums of independent, positive random variables and its applications to branching processes. Theory Probab. Appl. 9:640–648
36. Cramér H (1930) On the Mathematical Theory of Risk. Skandia Jubilee Volume, Stockholm
37. Cramér H (1955) Collective Risk Theory. Skandia Jubilee Volume, Stockholm
38. Dahms R (2012) Linear stochastic reserving methods. ASTIN Bull. 42:1–34
39. Daley DJ, Vere-Jones D (2008) An Introduction to the Theory of Point Processes, 2nd edn. Springer-Verlag, New York
40. **Dassios, A.** (1987). Ph.D. Thesis. *Imperial College, London*
41. Dassios A, Embrechts P (1989) Martingales and insurance risk. Commun. Statist. Stochastic Models 5:181–217
42. Dellacherie C, Meyer PA (1980) Probabilités et Potentiel. Ch, VI, Hermann, Paris
43. Deprez O, Gerber H (1985) On convex principles of premium calculation. Insurance Math. Econom. 4:179–189
44. **de Vylder, F.** (1978). A practical solution to the problem of ultimate ruin probability. *Scand. Actuarial J.*, 114–119
45. Dhaene J, Goovaerts MJ, Kaas R (2003) Economic capital allocation derived from risk measures. North Amer. Actuarial J. 7(2):44–56
46. Dickson DCM (1992) On the distribution of the surplus prior to ruin. Insurance Math. Econom. 11:191–207
47. Dickson DCM (2005) Insurance Risk and Ruin. Cambridge University Press, Cambridge
48. Doney RA (1991) Hitting probabilities for spectrally positive Lévy processes. J. London Math. Soc. 2(44):566–576
49. Dufresne F, Gerber HU (1988) The surpluses immediately before and at ruin, and the amount of the claim causing ruin. Insurance Math. Econom. 7:193–199
50. **DuMouchel, W.** (1968). The Pareto optimality of an n-company reinsurance treaty. *Scand. Actuarial J.*, 165–170
51. Eisenberg J, Schmidli H (2009) Optimal Control of Capital Injections by Reinsurance in a Diffusion Approximation. Blätter der DGVFM 30:1–13
52. **Eisenberg, J.** and **Schmidli, H.** (2011). Minimising Expected Discounted Capital Injections by Reinsurance in a Classical Risk Model. *Scand. Actuarial J.*, 155–176
53. Elliott MW, Webb BL, Anderson HN, Kensicki PR (1997) Principles of Reinsurance. Insurance Institute of America, Malvern, PA

54. Embrechts P, Goldie CM, Veraverbeke N (1979) Subexponentiality and infinite divisibility. Z. Wahrsch. verw. Geb. 49:335–347
55. **Embrechts, P., Grandell, J.** and **Schmidli, H.** (1993). Finite-time Lundberg inequalities in the Cox case. *Scand. Actuarial J.,* 17–41
56. Embrechts P, Veraverbeke N (1982) Estimates for the probability of ruin with special emphasis on the possibility of large claims. Insurance Math. Econom. 1:55–72
57. England P, Verall R (2002) Stochastic claims reserving in general insurance. Brit. Act. J. 8:443–544
58. Ethier SN, Kurtz TG (1986) Markov Processes. Wiley, New York
59. Feller W (1971) An Introduction to Probability Theory and its Applications, vol II. Wiley, New York
60. Ferguson TS (1972) A Bayesian analysis of some non-parametric problems. Ann. Statist. 1:209–230
61. Föllmer H, Schied A (2002) Convex measures of risk and trading constraints. Finance Stoch. 6:429–447
62. Gerber HU (1973) Martingales in risk theory. Schweiz. Verein. Versicherungsmath. Mitt. 73:205–216
63. Gerber HU (1975) The surplus process as a fair game - utilitywise. ASTIN Bull. 8:307–322
64. Gerber HU (1978) Pareto-optimal risk exchanges and related decision problems. ASTIN Bull. 10:25–33
65. Gerber HU (1979) An Introduction to Mathematical Risk Theory. Huebner Foundation Monographs, Philadelphia
66. Gerber HU, Goovaerts MJ, Kaas R (1987) On the probability and severity of ruin. ASTIN Bull. 17:151–163
67. Gerber HU, Jones DA (1975) Credibility formulas of the updating type. Trans. Soc. Actuaries 27:31–46
68. Gerber HU, Shiu ESW (1998) On the time value of ruin. North Amer. Actuarial J. 2(1):48–76
69. Glynn PW, Iglehart DL (1989) Importance sampling for stochastic simulation. Managment Science 35:1367–1392
70. Goelden H-W, Hess KT, Morlock M, Schmidt KD, Schröter KJ (2016) Schadenversicherungsmathematik. Springer, Heidelberg
71. Goovaerts MJ, Kaas R, van Heerwaarden AE, Bauwelinckx T (1990) Effective Actuarial Methods. North-Holland, Amsterdam
72. Goovaerts MJ, de Vylder F, Haezendonck J (1984) Insurance Premiums. North Holland, Amsterdam
73. **Grandell, J.** (1977). A class of approximations of ruin probabilities. *Scand. Actuarial J.,* 37–52
74. Grandell J (1991) Aspects of Risk Theory. Springer-Verlag, New York
75. Grandell J (1991) Finite time ruin probabilities and martingales. Informatica. 2:3–32
76. Grandell J (1995) Some remarks on the Ammeter risk process Schweiz. Verein. Versicherungsmath. Mitt. 95:43–71
77. Grandell J (1997) Mixed Poisson Processes. Chapman & Hall, London
78. Gray RJ, Pitts SM (2012) Risk Modelling in General Insurance. Cambridge University Press, Cambridge
79. Grenander U (1957) Some remarks on bonus systems in automobile insurance. Scand. Actuarial J. 40:180–198
80. Gut A (2009) Stopped Random Walks, 2nd edn. Springer-Verlag, New York
81. **Hachemeister, C.A.** and **Stanard, J.N.** (1975). IBNR claims count estimation with static lag functions. *12th Astin Colloquium, IAA, Portimao, Portugal*
82. Hadwiger H (1940) Über die Wahrscheinlichkeit des Ruins bei einer grossen Zahl von Geschäften. Archiv für mathematische Wirtschafts- und Sozialforschung 6:131–135
83. Hald M, Schmidli H (2004) On the maximisation of the adjustment coefficient under proportional reinsurance. ASTIN Bull. 34:75–83

84. **Hesselager, O.** (1990). Some results on optimal reinsurance in terms of the adjustment coefficient. *Scand. Actuarial J.*, 80–95
85. **Hipp, C.** and **Schmidli, H.** (2004). Asymptotics of ruin probabilities for controlled risk processes in the small claims case. *Scand. Actuarial J.*, 321–335
86. Iglehart DL (1969) Diffusion approximations in collective risk theory. J. Appl. Probab. 6:285–292
87. Janssen J (1980) Some transient results on the M/SM/1 special semi-Markov model in risk and queueing theories. ASTIN Bull. 11:41–51
88. Kaas R, Goovaerts M, Dhaene J, Denuit M (2002) Modern Actuarial Risk Theory. Kluwer Academic Publishers, Dordrecht
89. Karlin S, Taylor HM (1975) A First Course in Stochastic Processes. Academic Press, New York
90. Karlin S, Taylor HM (1981) A Second Course in Stochastic Processes. Academic Press, New York
91. Klugman SA, Panjer HH, Willmot GE (2012) Loss Models. From data to decisions. Fourth edition, Wiley Hoboken, NJ
92. Klüppelberg C (1988) Subexponential distributions and integrated tails. J. Appl. Probab. 25:132–141
93. Kremer E (1985) Einführung in die Versicherungsmathematik. Vandenhoek & Ruprecht, Göttingen
94. Kyprianou AE (2006) Introductory lectures on fluctuations of Lévy processes with applications. Springer-Verlag, Berlin
95. Lemaire J (1995) Bonus-Malus Systems in Automobile Insurance. Kluwer Academic Publishers, Boston
96. **Lundberg, F.** (1903). *I. Approximerad Framställning av Sannolikhetsfunktionen. II. Återförsäkring av Kollektivrisker.* Almqvist & Wiksell, Uppsala
97. Lundberg F (1926) Försäkringsteknisk Riskutjämning. F. Englunds boktryckeri A.B, Stockholm
98. Mack T (1991) A simple parametric model for rating automobile insurance or estimating IBNR claims reserves. ASTIN Bull. 21:93–109
99. Mack T (1993) Distribution-free calculation of the standard error of chain ladder reserve estimates. ASTIN Bull. 23:213–225
100. Mack T (1997) Schadenversicherungsmathematik. Verlag Versicherungswirtschaft, Karlsruhe
101. Mack T (2000) Credible claims reserves: the Benktander method. ASTIN Bull. 30:333–347
102. **Neuhaus, W.** (1992). Another pragmatic loss reserving method or Bornhuetter/Ferguson revisited. *Scand. Actuarial J.*, 151–162
103. **Norberg, R.** (1979). The credibility approach to experience rating. *Scand. Actuarial J.*, 181–221
104. Norberg R (1999) Prediction of outstanding liabilities II. ASTIN Bull. 29:5–25
105. Ohlin J (1969) On a class of measures of dispersion with application to optimal reinsurance. ASTIN Bull. 5:249–266
106. Panjer HH (1981) Recursive evaluation of a family of compound distributions. ASTIN Bull. 12:22–26
107. **Panjer, H.H.** and **Willmot, G.E.** (1992). *Insurance Risk Models.* Society of Actuaries
108. Radtke M, Schmidt KD, Schnaus A (2016) Handbook on Loss Reserving. Springer International Publishing, Switzerland
109. Röhr A (2016) Chain ladder and error propagation. ASTIN Bull. 46:293–330
110. Rolski T, Schmidli H, Schmidt V, Teugels JL (1999) Stochastic Processes for Insurance and Finance. Wiley, Chichester
111. Ross SM (1996) Stochastic Processes, 2nd edn. Wiley, New York
112. **Schmidli, H.** (1992). A general insurance risk model. *Ph.D Thesis, ETH Zürich*
113. Schmidli H (1994) Diffusion approximations for a risk process with the possibility of borrowing and investment. Comm. Statist. Stochastic Models 10:365–388

114. Schmidli H (1997) An extension to the renewal theorem and an application to risk theory. Ann. Appl. Probab. 7:121–133
115. Schmidli H (1999) On the distribution of the surplus prior and at ruin. ASTIN Bull. 29:227–244
116. Schmidli H (2008) Stochastic Control in Insurance. Springer-Verlag, London
117. Schmidli H (2014) Extended Gerber-Shiu functions in a risk model with interest. Insurance Math. Econom. 61:271–275
118. Schnieper R (1990) Insurance premiums, the insurance market and the need for reinsurance. Schweiz. Verein. Versicherungsmath. Mitt. 90:129–147
119. Seal HL (1972) Numerical calculation of the probability of ruin in the Poisson/exponential case. Schweiz. Verein. Versicherungsmath. Mitt. 72:77–100
120. **Seal, H.L.** (1974). The numerical calculation of $U(w, t)$, the probability of non-ruin in an interval $(0, t)$. *Scand. Actuarial J.*, 121-139
121. **Segerdahl, C.-O.** (1955). When does ruin occur in the collective theory of risk? *Skand. Aktuar Tidskr.*, 22–36
122. Siegmund D (1979) Corrected diffusion approximation in certain random walk problems. Adv. in Appl. Probab. 11:701–719
123. Sundt B (1999) An Introduction to Non-Life Insurance Mathematics. Verlag Versicherungswirtschaft e.V, Karlsruhe
124. Stam AJ (1973) Regular variation of the tail of a subordinated probability distribution. Adv. Appl. Probab. 5:308–327
125. Takács L (1962) Introduction to the Theory of Queues. Oxford University Press, New York
126. Tarbell TF (1934) Incurred but not reported claim reserves. Proc. CAS 20:275–280
127. Taylor G (2000) Loss Reserving: An Actuarial Perspective. Kluwer Academic Publishers, Boston
128. Taylor G (2015) Bayesian chain ladder models. ASTIN Bull. 45:75–99
129. **Thorin, O.** (1974). On the asymptotic behavior of the ruin probability for an infinite period when the epochs of claims form a renewal process. *Scand. Actuarial J.*, 81–99
130. **Thorin, O.** (1975). Stationarity aspects of the Sparre Andersen risk process and the corresponding ruin probabilities. *Scand. Actuarial J.*, 87–98
131. **Thorin, O.** (1982). Probabilities of ruin. *Scand. Actuarial J.*, 65–102
132. Thorin O, Wikstad N (1977) Calculation of ruin probabilities when the claim distribution is lognormal. ASTIN Bull. 9:231–246
133. **von Bahr, B.** (1975). Asymptotic ruin probabilities when exponential moments do not exist. *Scand. Actuarial J.*, 6–10
134. von Neumann J, Morgenstern O (1944) Theory of Games and Economic Behavior. Princeton University Press, Princeton
135. Wang SS (1996) Premium calculation by transforming the layer premium density. ASTIN Bull. 26:71–92
136. Waters HR (1983) Some mathematical aspects of reinsurance insurance. Insurance Math. Econom. 2:17–26
137. Whitney AW (1918) The theory of experience rating. Proc. Caualty Actuar. Soc. 4:274–292
138. Whitt W (1970) Weak convergence of probability measures on the function space $C[0, \infty)$. Ann. Statist. 41:939–944
139. **Williams, D.** (1979). *Diffusions, Markov Processes and Martingales.* Volume I, Wiley, New York
140. Wüthrich MV, Merz M (2008) Stochastic Claims Reserving Methods in Insurance. Wiley, Chichester
141. **Wüthrich, M.V.** and **Merz, M.** (2015). Stochastic Claims Reserving Manual: Advances in Dynamic Modeling. *SSRN, http://ssrn.com/abstract=2649057*

Index

H. Schmidli, *Risk Theory*, Springer Actuarial,
DOI:10.1007/978-3-319-72005-0

Made in United States
Orlando, FL
14 September 2026

87736175R00142